IMPROVING FLOOD PREDICTION ASSIMILATING UNCERTAIN CROWDSOURCED DATA INTO HYDROLOGIC AND HYDRAULIC MODELS

IMPROVING FLOOD PREDICTION ASSIMILATING UNCERTAIN CROWDSOURCED DATA INTO HYDROLOGICAL AND HYDRAULIC MODELS

DISSERTATION

Submitted in fulfilment of the requirements of
the Board for Doctorates of Delft University of Technology
and of
the Academic Board of the UNESCO-IHE Institute for Water Education
for the Degree of DOCTOR
to be defended in public on
Monday, 28 November 2016, at 10:00 hours
In Delft, the Netherlands

by

Maurizio MAZZOLENI
Master of Science in Environmental Engineering
University of Brescia, Brescia

born in Brescia, Italy

This dissertation has been approved by the

promotor: Prof. dr. D.P. Solomatine and
copromotor: Dr. L. Alfonso

Composition of the doctoral committee:

Chairman	Rector Magnificus TU Delft
Vice-Chairman	Rector UNESCO-IHE
Prof.dr. D.P. Solomatine,	UNESCO-IHE/TU Delft, promotor
Dr. L. Alfonso,	UNESCO-IHE, copromotor

Independent members:

Prof. dr. ir. A.W. Heemink,	TU Delft
Prof. dr. ir. A. Weerts,	Wageningen University
Dr. ir. H. Madsen,	University of Copenhagen, Denmark
Prof. dr. ir J.A. Roelvink,	UNESCO-IHE/TU Delft
Prof. dr. N.C. van de Giesen,	TU Delft, reserve

This research was conducted under the auspices of the Graduate School for Socio-Economic and Natural Sciences of the Environment (SENSE)

CRC Press/Balkema is an imprint of the Taylor & Francis Group, an informa business

Published by:
CRC Press/Balkema
PO Box 11320, 2301 EH Leiden, The Netherlands
Pub.NL@taylorandfrancis.com
www.crcpress.com – www.taylorandfrancis.com

ISBN 978-1-138-03590-4 (Taylor & Francis Group)

To my parents, Mirella and Luciano

To Jessica

ACKNOWLEDGMENTS

Once Nelson Mandela said "It always seems impossible until it's done". Indeed, during my PhD research I had this feeling many times, but it is because of the help of my supervisors, colleagues, family and friends that I have finally made it.

First of all, I would like to express my sincere gratitude to my promotor Prof. Dimitri Solomatine and co-promotor Leonardo Alfonso. Dear Dimitri, in these four years you provided me with creative and innovative ideas, critical thought and advice that inspired me during my PhD research. Thanks for always believing in me, and involving me in numerous research activities and projects. Leo, thank you for always being available for a nice word or suggestion, for sharing your ideas and for the insightful discussions on innovative research topics.

This PhD research project was part of the WeSenseIt Project which was funded by the Seventh Framework Programme of the European Union. I really enjoyed being part of such an important project and I would like to thank all the partner for their support. In particular, I appreciate the dedication of Martina, Daniele and Michele of the Alto Adriatico Water Authority for sharing flood data together with their hydrological and hydraulic models with us.

I am grateful to Jolanda, Anique and Jos for all the administrative support, Paula Derkse for the suggestion for printing this thesis, Benno for helping me with the samenvatting and Joanne for the proofread of the thesis. Thanks to all the Hydroinformatics staff members, Ioana, Andreja, Schalk Jan, Biswa, and Gerald for involving me in the teaching activities of our group.

I am very thankful to Prof. Dong-Jun Seo for the opportunity to spend four months as a visiting scholar at the University of Texas at Arlington where I also met Dr. Seong-Jin Noh. Dear D.J. and Seong-Jin, thank you very much for your hospitality and your precious scientific advice on data assimilation, I'm sure in the future there will be opportunities to strengthen the collaboration created between UTA and UNESCO-IHE. Special thanks go to Behzad Z., Hamideh R., Hamideh H., Ray and Bahzad N. for making me feel at home in Arlington.

Luigia, thank you very much for always being there when I needed it, for your sincere comments on the readability of our papers and for the funny moments during the coffee breaks. Thank you Giuliano for being not only a supervisor but also a friend. It is also because of you if I am now here. I wish to you, Luigia and Tommaso all the best.

There are no words to convey my feelings to all my friends in Delft. Thanks Yared for the 3 years sharing the workspace, for the Saturday football in TU, for all the fun during conferences and many other moments. Juan, project buddy, thanks for all the discussions about data assimilation, for the support during the PhD and for helping to finish all that vodka in Krakow. Kun, sometimes it was more difficult to explain my jokes to you than data assimilation to my mother, thanks for all those nice memories. Thanks Angi, my friiiend, for all the fonny moments and your patience with me in the gym. Thanks Maribel, I could not ask for a better roommate, even if we had some bread-related issues. Thank you to all of you guys, Benno, Elena, Paolo, Jojo, Juliette, Aki, Pato, Miguel, Alex, Alessio, Alessandro, Neiler, Vero, Mark, Fer, Peter, Alida, Pablo, Claudia, Sara, Arlex, ShanShan, Erika, Andres, Juan Pablo, Paola, Natalia, Micah, Mario, Quan, Mohaned, Stefan, Quintilia, Dibesh, Joanne, Imra, Hadi, Feroz, Marianne, Laura, Victor, and many more for being here all this time.

I would like to express my immense gratitude to Vale and Matteo for their continuous support, Ale and Livio for unforgettable moments in the mountains and inspiring conversations and Davide and Michela for their Sunday pizza dinners and wedding talks. I would like to thank also Cristian, Manuela, Ippolita, Gianni, Marco, Ciro, Manuel and Laura for always finding time for me when I was back in Italy even if very busy.

Last but not least, I would like to give a special thanks to someone who has been constantly with me for these last four years, in the rainy and sunny days, in the sad and happy moments and, many times, overnight. We had difficult moments but at the end, somehow, it always "ran" smoothly. Thank you my dear Laptop.

Un ringraziamento speciale va ai miei genitori, Mirella e Luciano. Grazie per la vostra fiducia, paziena e per avermi sempre sostenuto incondizionatamente in questi ultimi quattro anni (e non solo) anche quando le mie scelte mi hanno portato lontano da casa per molto tempo. Un pezzo di questo dottorato é anche vostro. Grazie di cuore ai miei zii Rossella, Battista e Gianni per essere sempre stati li quando c'era bisogno e per aver aiutato la mamma negli ultimi due anni. Grazie Daniela per aver supportato ma soprattutto sopportato quel vecchio brontolone in questi ultimi anni.

I do not think I would ever get to finish my PhD without you, my wife, my lover, my best friend, Jessica. In these last four years you saw the best and worst parts of me and you were always there to cheer me up. My journey would not have been possible without your continuous and unlimited support and help. Thanks for your love, encouraging me in all of my pursuits and inspiring me to follow my dreams. Te amo

Summary

Monitoring stations have been used for decades to measure hydrological variables, and mathematical water models used to predict floods can be enhanced by the incorporation of these observations, i.e. by data assimilation. The assimilation of remotely sensed water level observations in hydrological and hydraulic modelling has become more attractive due to their availability and spatially distributed nature.

In recent years, continued technological advances have stimulated a spread of low-cost sensors that has triggered crowdsourcing as a way to obtain observations of hydrological variables in a more distributed way than the classic static physical sensor networks. The main advantage of using this type of sensors is that they can be used not only by technicians, as with observations from traditional physical sensors, but also by regular citizens. However, there are also drawbacks of using these observations, e.g. their relatively limited reliability, varying accuracy in time and space, and their irregular and non a-priori defined availability. For this reason, crowdsourced observations have not been widely integrated in hydrological and/or hydraulic models for flood forecasting applications. Instead, they have generally been used to validate model results against observations, in post-event analyses.

Model updating is a strategy that aims at improving models using observations. A particular case of model updating is data assimilation, which often uses measured data such as streamflow, soil moisture, etc. coming from static physical stations. However, only a few studies have considered the integration of crowdsourced observations into water-related models.

The main objective of this research is to investigate the benefits of assimilating crowdsourced observations coming from a distributed network of heterogeneous physical and social (static and dynamic) sensors within hydrological and hydraulic models, in order to improve flood forecasting. Standard data assimilation approaches, such as Kalman filtering, ensemble Kalman filtering, nudging, etc. are applied to the three different case studies to assimilate crowdsourced observations of variable accuracy and random life-span. The results of this study demonstrate that crowdsourced observations can significantly improve flood prediction if they are properly integrated in hydrological and hydraulic models.

In particular, this research proved that assimilation of streamflow observations from static physical sensors provides improvements in model performance, the magnitude of which depends on the observation locations and model structure. In case of the Brue catchment, the best model improvement is achieved by assimilating streamflow observation along the main river reach. However, the varying spatial distribution of precipitation generating flood events affects which sensor locations produce the best model performance at the catchment outlet.

This research also demonstrated that assimilation of crowdsourced streamflow observations at interior points of the catchment can improve model performances, dependent upon the particular location of the static social sensors and the hydrological model structures. In this case, lower accuracy, variable in time and space, is assumed for crowdsourced data from social sensors than for physical sensors. In case of the Brue catchment, realistic assumptions about the locations (next to urban areas where citizens can provide data) and temporal availability (mainly daylight hours) of crowdsourced observations from static social sensors, optimally and non-optimally located, are introduced. Interestingly, it is demonstrated that hydrological models can perform better with appropriately distributed social sensors than with inappropriately distributed physical sensors. For this reason, a non-optimal distribution of static physical sensors can be integrated with a network of static social sensors, providing intermittent crowdsourced observations in order to improve model performance.

Citizen-based crowdsourced observations are generally characterised by random accuracy and are derived at random (asynchronous) moments, which may not coincide with the model time step. The results of this research show that, for a given sensor location, there is a limit to the number of assimilated crowdsourced asynchronous observations, after which only marginal model improvements are obtained. Accuracy of the crowdsourced observations influences the model results more than the time of arrival of the data. Nash-Sutcliffe index values drop when the intervals between the assimilated observations are too large. In this case, the abundance of crowdsourced data is no longer able to compensate their intermittency. In experiments with the Bacchiglione catchment it is proved that a single physical sensor can be complemented with distributed static social sensors providing asynchronous observations, even with a limited number of intermittent asynchronous crowdsourced measurements.

Regarding hydraulic modelling, different data assimilation approaches (such as direct insertion, nudging, Kalman filtering, ensemble Kalman filtering and asynchronous ensemble Kalman filtering) are implemented to integrate streamflow and water depth observations from static social and physical sensors at different locations. In general, assimilation of streamflow observations in both lumped and distributed structures of a 3-parameter Muskingum model increases model performance. Furthermore, it is found that direct insertion works better for lumped models, while ensemble Kalman filtering approaches are more reliable for distributed models. This can be due to the fact that using direct insertion model states are updated only at the assimilation location, while using Kalman filtering approaches the update is performed along the whole river reach because of the distributed nature of the Kalman gain and covariance matrix. Increasing the number of past observations in the asynchronous ensemble Kalman filter improves model performance expressed in terms of Nash-Sutcliffe, correlation and Bias indexes. Nonetheless, Kalman filtering methods are noticeably sensitive to the degree of model error and sensor locations.

Considering assimilation of distributed water depth observations in a linear hydraulic model, such as a Muskingum-Cunge model, it is found that the Kalman filter is noticeably sensitive to the degree of model error and sensor location. When the largest error is found in the boundary condition, the optimal sensor location is close to the boundary condition point, while if the error in the model exceeds the boundary condition error, the optimal sensor location is close to the reach outlet. In the Bacchiglione River it is shown that assimilating water depth observations from reaches close to the river outlet rather than from upstream reaches tends to provide larger improvement. However, downstream reaches tend to lose the assimilation effects faster than the upstream ones, in the case of flood prediction, due to their shorter travel time. In consequence, the optimal location of static physical and social sensors should be considered as a compromise between the largest model improvement and the prediction capability of the model itself.

Finally, water depth observations from distributed physical (static) and social (static and dynamic) sensors are assimilated within the semi-distributed hydrological and hydraulic model of the Bacchiglione catchment. Assimilation of crowdsourced data into hydrological models led to good model predictions for long lead times. On the other hand, assimilation of crowdsourced observations in river reaches close to the catchment outlet guarantees the best model prediction for short lead times. In addition, different observation bias scenarios are considered. In the case of realistic scenarios of different citizen engagement levels, this research shows that sharing crowdsourced observations motivated by a feeling of belonging to a community helps in improving flood predictions. In particular, the model results can benefit from the additional observations provided by weather enthusiasts.

This research demonstrates that networks of low-cost static and dynamic social sensors can complement traditional networks of static physical sensors, for the purpose of improving flood forecasting accuracy. This can be a potential application of recent efforts to build citizen observatories of water, in which citizens not only can play an active role in information capturing, evaluation and communication, but can also help to improve models and thus increase flood resilience.

Maurizio Mazzoleni

Delft, the Netherlands

SAMENVATTING

Meetstations zijn de afgelopen decennia gebruikt om hydrologische variabelen te meten. Mathematische watermodellen die gebruikt worden om overstromingen te voorspellen, kunnen verbeterd worden door de inclusie van deze metingen, bijvoorbeeld door data-assimilatie. De assimilatie van op satelliet-afstand waargenomen metingen van het waterniveau in hydrologische en hydraulische modellering is aantrekkelijker geworden vanwege de beschikbaarheid en ruimtelijke distributie.

Voortschrijdende technologische ontwikkelingen hebben de afgelopen jaren de verspreiding van laaggeprijsde sensoren gestimuleerd. Dit heeft geleid tot crowdsourcing: een manier om hydrologische variabelen in een meer wijdverspreide manier te observeren dan de klassieke statische, fysieke sensornetwerken. Het grootste voordeel van dit soort sensoren, is dat deze niet alleen door technici gebruikt kunnen worden, zoals het geval is met metingen van traditionele fysieke sensoren, maar ook door gewone burgers. Echter, er zijn ook nadelen bij het gebruik van deze metingen; bijvoorbeeld de relatief beperkte betrouwbaarheid, wisselende nauwkeurigheid in tijd en ruimte en de willekeurige en niet vooraf gedefinieerde beschikbaarheid. Vanwege deze redenen zijn crowdsourced metingen niet breed geïntegreerd in hydrologische en/of hydraulische modellen voor applicaties voor het voorspellen van overstromingen. In plaats daarvan worden ze meestal gebruikt om modellen te valideren ten opzichte van metingen in post-event analyses.

Model updating is een strategie die gericht is op het verbeteren van modellen met behulp van metingen. Een bijzonder geval van model updating is data-assimilatie, welke vaak gebruikt maakt van informatie van de stroming in rivieren, bodemvocht, etc. afkomstig van statische fysieke stations. Echter, slechts een paar studies hebben de integratie van crowdsourced metingen in watergerelateerde modellen in beschouwing genomen.

Het belangrijkste doel van dit onderzoek is om de voordelen van het assimileren van de crowdsourced waarnemingen, afkomstig van een gedistribueerd netwerk van heterogene fysieke en sociale (statische en dynamische) sensoren binnen hydrologische en hydraulische modellen te onderzoeken, om zo het voorspellen van overstromingen te verbeteren. Standaard data-assimilatiebenaderingen, zoals Kalman filtering, ensemble Kalman filtering, nudging, etc. worden toegepast op de drie verschillende case studies om crowdsourced waarnemingen met een variabele nauwkeurigheid en willekeurige levensduur te assimileren. De resultaten van deze studie tonen aan dat crowdsourced waarnemingen de voorspelling van overstromingen significant kunnen verbeteren, wanneer deze goed in hydrologische en hydraulische modellen geïntegreerd zijn.

Dit onderzoek heeft in het bijzonder bewezen dat assimilatie van stromingsmetingen van statische, fysieke sensoren verbeteringen brengt in de modelprestaties. De mate waarin dit het geval is, hangt af van de locaties van de metingen en de modelstructuur. In het geval van het Brue stroomgebied is de beste modelverbetering bereikt door het assimileren van streamflow metingen langs belangrijkste rivierbereik. Echter, de variërende ruimtelijke verdeling van de neerslag-genererende overstromingsgebeurtenissen heeft invloed op welke sensorlocaties de beste modelprestaties op het einde van het stroomgebied hebben.

Het onderzoek toonde ook aan dat de assimilatie van crowdsourced stromingsmetingen op de binnenste punten van het stroomgebied de modelprestaties kunnen verbeteren, afhankelijk van de bijzondere ligging van de statische sociale sensoren en de hydrologisch modelstructuren. In dit geval wordt een lagere nauwkeurigheid, variabel in tijd en ruimte, verondersteld bij crowdsourced data van sociale sensoren in vergelijking met die van fysieke sensoren. In het geval van het Brue stroomgebied worden realistische veronderstellingen over de locaties (naast de stedelijke gebieden waar burgers de data kunnen leveren) en tijdelijke beschikbaarheid (voornamelijk bij daglicht) van crowdsourced metingen van optimaal en niet-optimaal geplaatste statische sociale sensoren geïntroduceerd. Interessant is dat het aangetoond is dat hydrologische modellen beter kunnen presteren met zorgvuldig gedistribueerde sociale sensoren dan met meer willekeurig gedistribueerde fysieke sensoren. Daarom kan een niet-optimale verdeling van statische fysieke sensoren worden geïntegreerd met een netwerk van statische sociale sensoren, leidend tot intermitterende crowdsourced metingen voor modelverbetering.

Crowdsource waarnemingen door burgers worden in het algemeen gekenmerkt door willekeurige nauwkeurigheid en zijn afgeleid op willekeurige (asynchrone) momenten, die mogelijk niet samenvallen met de modeltijdstap. De resultaten van dit onderzoek wijzen erop dat, bij op een bepaalde sensorlocatie, er een limiet geldt voor het aantal geassimileerde crowdsourced asynchrone metingen, waarna er slechts marginale modelverbeteringen plaatsvinden. Nauwkeurigheid van de crowdsourced metingen beïnvloed de modelresultaten meer dan het moment waarop de data arriveert. Nash-Sutcliffe index waarden nemen af wanneer de intervallen tussen de geassimileerd metingen te groot worden. In dit geval kan de overvloed aan crowdsourced data niet meer de intermitterende aard compenseren. In experimenten met het Bacchiglione stroomgebied wordt aangetoond dat een enkele fysieke sensor kan worden aangevuld met gedistribueerde statische sociale sensoren die asynchrone metingen verschaffen, zelfs bij een beperkt aantal intermitterende asynchrone crowdsourced metingen.

Ten aanzien van de hydraulische modellering worden verschillende data-assimilatie benaderingen (zoals direct inbrengen, nudging, Kalman filtering, ensemble Kalman filtering en asynchrone ensemble Kalman filtering) geïmplementeerd om rivierstroming- en waterdieptemetingen te integreren vanuit statische sociale en fysieke sensoren op verschillende locaties. Over het algemeen verbetert assimilatie van de stromingsmetingen in zowel gecombineerde als gedistribueerde structuren van een 3-parameter Muskingum model de prestatie van het model. Bovendien blijkt dat directe insertie beter werkt voor gecombineerde modellen, terwijl ensemble Kalman filtering benaderingen betrouwbaarder zijn voor gedistribueerde modellen. Dit kan te wijten zijn aan het feit dat bij directe insertie modellen alleen worden geüpdatet op de assimilatieplaats, terwijl bij Kalman filteringsbenaderingen de update uitgevoerd wordt langs het hele rivierbereik vanwege het gedistribueerde karakter van het Kalman gain en covariantie matrix. Verhoging van het aantal waarnemingen uit het verleden in de asynchrone ensemble Kalman filter verhoogt de modelprestatie uitgedrukt in termen van Nash-Sutcliffe, correlatie en Bias indexen. Niettemin, Kalman filtering methoden zijn merkbaar gevoelig voor de waarde van modelfout en sensorlocaties.

In het geval van assimilatie van gedistribueerde waterdieptemetingen in een lineair hydraulisch model, zoals Muskingum-Cunge model, blijkt dat het Kalma filter merkbaar gevoelig voor de waarde van de modelfout en de sensorlocatie. Als de grootste fout gevonden wordt in de randconditie, dan is de optimale sensorlocatie dichtbij het punt van de randvoorwaarde, terwijl als de fout in het model groter is dan de fout van de randvoorwaarde, dan is de optimale sensorlocatie dichtbij het einde van het rivierbereik. In de rivier Bacchiglione is aangetoond dat assimilatie van de waterdieptewaarnemingen van bereik in de buurt van het riviereinde grotere verbetering kan bieden in vergelijking stroomopwaarts bereik. Echter, stroomafwaarts bereik neigt vaak de assimilatie-effecten sneller te verliezen dan stroomopwaarts bij vloedvoorspelling, door hun kortere reistijd. Bijgevolg is dat de optimale locatie van statisch fysieke en sociale sensoren zou moeten worden beschouwd als een compromis tussen de grootste modelverbetering en het voorspellend vermogen van het model zelf.

Tot slot, waterdieptemetingen van gedistribueerde fysieke (statische) en sociale (statische en dynamische) sensoren zijn geassimileerd in het semi-gedistribueerde hydrologische en hydraulische model van het Bacchiglione stroomgebied. Assimilatie van crowdsourced gegevens in hydrologische modellen zorgde voor een goede modelvoorspellingen met lange doorlooptijd. Aan de andere kant staat assimilatie van crowdsourced waarnemingen in de river in de buurt van het einde van het stroomgebied garant voor de beste modelprestaties met korte doorlooptijd. Aanvullend worden verschillende bias scenario's beschouwd. In het geval van realistische scenario's van verschillende betrokkenheidniveaus van burgers, toont dit onderzoek aan dat het delen van crowdsourced metingen helpt bij het verbeteren van overstromingsvoorspellingen, gemotiveerd door een gevoel van betrokkenheid bij de gemeenschap. In het bijzonder kunnen de modelresultaten profiteren van de aanvullende metingen gedaan door liefhebbers van meteorologie.

Dit onderzoek toont aan dat netwerken van laaggeprijsde statische en dynamische sociale sensoren een aanvulling kunnen zijn op traditionele netwerken van statische fysieke sensoren, met betrekking tot het verbeteren van de nauwkeurigheid van het voorspellen van overstromingen. Dit kan leiden tot een mogelijke toepassing van de recente inspanningen om de waterobservatoria voor burgers te bouwen, waarin burgers niet alleen een actieve rol in spelen in het vastleggen van informatie, evaluatie en communicatie, maar ook kunnen helpen bij het verbeteren van modellen en daarbij de omgang met overstromingen.

Maurizio Mazzoleni

Delft, the Netherlands

SOMMARIO

Stazioni di monitoraggio sono state utilizzate per decenni per misurare correttamente variabili idrologiche e meglio prevedere le inondazioni. A tal fine, modelli matematici sono stati sviluppati e migliorati usando metodi che incorporano tali osservazioni idrologiche. L'assimilazione di livelli idrici da satellite in modelli idrologici ed idraulici è diventata sempre più attraente ed usata a causa della loro disponibilità e la natura spazialmente distribuita.

Negli ultimi anni, i continui progressi tecnologici hanno stimolato la diffusione di sensori a basso costo che ha innescato il fenomeno del crowdsourcing come un strumento per ottenere osservazioni di variabili idrologiche in modo più distribuito rispetto ai classici sensori fisici statici. Il principale vantaggio di utilizzare questo tipo di sensori è che possono essere utilizzati non solo da tecnici, come dei sensori fisici tradizionali, ma anche da regolari cittadini. Tuttavia, ci sono anche svantaggi legati all´uso di queste osservazioni in modelli matematici, quali per esempio loro limitata affidabilità, la precisione variabile in tempo e spazio, e la loro disponibilità irregolare. Per questo motivo, fino a questo momento, osservazioni crowdsourcing non sono state opportunamente integrate in modelli idrologici e/o idraulici al fine di migliorare la previsione di piena ma il loro utilizzo è stato focalizzato in analisi post evento per la convalida di modelli matematici.

L'obiettivo principale di questa tesi è quello di analizzare i benefici legati all´assimilazione di osservazioni crowdsourcing, provenienti da una rete eterogenea e distribuita di sensori fisici e sociali (statici e dinamici), all'interno di modelli idrologici ed idraulici, al fine di migliorare la previsione di piena. Approcci standard di assimilazione dei dati, come ad esempio il filtro di Kalman, sono stati applicati in tre diversi casi studio al fine di assimilare osservazioni crowdsourcing aventi precisione variabile in spazio e tempo. I risultati di questo studio dimostrano come le osservazioni crowdsourcing possano migliorare significativamente la previsione delle inondazioni, se opportunamente integrate in modelli idrologici ed idraulici.

In particolare, questa ricerca ha dimostrato che l'assimilazione di osservazioni di deflusso da sensori fisici statici fornisce miglioramenti nelle prestazioni del modello a seconda della posizione di tali osservazioni e struttura del modello matematico usato. Nel caso del bacino idrologico del fiume Brue, i migliori risultati sono stati raggiunti assimilando osservazioni di deflusso lungo il tratto fluviale principale.

Questa tesi ha dimostrato, inoltre, come l'assimilazione delle osservazioni crowdsourcing di deflusso nei punti interni del bacino sia in grado di migliorare le prestazioni del modello in base alla particolare posizione dei sensori statici sociali ed alla struttura del modello idrologico. In questo caso minore accuratezza, variabile nel tempo e nello spazio, è stata assunta per le osservazioni provenienti da sensori sociali rispetto a quella dei sensori fisici. Nel caso del bacino del fiume Brue, ipotesi realistiche sulla posizione (vicina alle aree urbane in cui i cittadini sono in grado di fornire i dati) e la disponibilità temporale (principalmente durante le ore diurne) di osservazioni crowdsourcing provenienti da sensori sociali statici e sensori fisici statici, posizionati in modo ottimale e non, sono state introdotte. È interessante notare come i modelli idrologici forniscano risultati migliori con sensori sociali opportunamente distribuiti rispetto ai sensori fisici impropriamente distribuiti. Per questo motivo, una rete non ottimale dei sensori fisici statici può essere integrata con una rete di sensori sociali statici.

Osservazioni crowdsourcing fornite da cittadini sono generalmente caratterizzate da precisione casuale ed inviate in momenti non definiti a priori (asincroni), che potrebbero non coincidere con il passo temporale del modello matematico. Questo studio ha mostrato che vi è un limite nel numero di osservazioni crowdsourcing asincroni assimilate dopo il quale i miglioramenti del modello ottenuti sono marginali. Infatti l'accuratezza delle osservazioni crowdsourcing influenza i risultati più che l'arrivo casuale di tali dati. Le performance del modello diminuiscono nel momento in cui gli intervalli tra le osservazioni assimilate sono troppo grandi. In questo modo l'abbondanza di dati crowdsourcing non è più in grado di compensare la loro intermittenza. Gli esperimenti eseguiti sul bacino del fiume Bacchiglione hanno dimostrato come un singolo sensore fisico può essere integrato con sensori sociali statici distribuiti anche nel caso di un numero limitato di misurazioni crowdsourcing asincroni intermittenti.

Per quanto riguarda la modellazione idraulica, diversi approcci per l'assimilazione dei dati, come ad esempio l'inserimento diretto, schema nudging, filtro di Kalman, filtro di Kalman di insieme e filtro di Kalman di insieme asincrono sono stati implementati per integrare dati di livello e di deflusso da sensori fisici e sociali installati in luoghi diversi. In generale, l'assimilazione di osservazioni di deflusso in strutture distribuite e concentrate di un modello Muskingum a 3 parametri aumenta le prestazioni del modello. Inoltre, è stato constatato come il metodo di inserimento diretto funziona meglio per i modelli a parametri concentrati, mentre gli approcci di filtraggio di Kalman di insieme sono più affidabili in caso di modelli distribuiti. Questo può essere dovuto al fatto che nel caso di inserzione diretta, gli stati del modello vengono aggiornati solo alla posizione di assimilazione, mentre nel caso di metodi Kalman l'aggiornamento viene eseguito lungo l'intero tratto del fiume a causa della natura distribuita del metodo. Aumentare il numero di osservazioni passate nel filtro di Kalman di insieme asincrono aumenta le prestazioni del modello stesso espresse in termini di indici di Nash-Sutcliffe, correlazione e Bias. Tuttavia, i metodi di filtraggio Kalman sono notevolmente sensibili all'errore del modello ed alla posizione dei sensori.

In caso di assimilazione di dati di livello distribuiti in un modello idraulico lineare, come ad esempio il modello Muskingum-Cunge, è stato constatato come il filtro di Kalman sia notevolmente sensibile al valore di errore del modello ed alla posizione del sensore. Un errore elevato nelle condizioni al contorno tende a migliorare miglior il profilo di corrente quando il punto di assimilazione è vicino ad esse, mentre in caso di errore di modello maggiore di quello delle condizioni al contorno buone prestazioni modello sono state ottenute se il sensore è situato vicino alla sezione di chiusura del tratto di fiume. Nel caso del fiume Bacchiglione è stato dimostrato che l'assimilazione di osservazioni di livello idrico in tratti di fiume vicino alla sezione di chiusura del fiume stesso (quindi a valle) tende a fornire un miglioramento superiore rispetto all'assimilazioni di dati a monte. Tuttavia, assimilazione in tratti a valle tende a far perdere gli effetti di tale assimilazione più velocemente di quelle a monte, in caso di previsione di piena. Di conseguenza, la posizione ottimale di sensori fisici e sociali statici dovrebbe essere considerato come un compromesso tra il miglior miglioramento del modello e la capacità di previsione del modello stesso.

Infine, osservazioni di livello idrico provenienti da sensori fisici (statici) e sociali (statici e dinamici) sono assimilate all'interno del modello semi-distribuito idrologico ed idraulico implementato su bacino del fiume Bacchiglione. L'assimilazione di dati crowdsourcing in bacini del modello idrologico garantisce buoni miglioramenti del modello stesso anche per i valori elevati di tempo di previsione. D'altra parte, l'assimilazione di osservazioni crowdsourcing nel modello idraulico del fiume vicino alla sezione di chiusura garantisce le migliori prestazioni del modello per i valori di tempo di previsione basso. In aggiunta, diverse situazioni di bias nelle osservazioni sono state considerate. Nel caso di realistici scenari di diversi livelli di coinvolgimento dei cittadini, questo studio ha dimostrato che condividere le osservazioni crowdsourcing, mossi da un sentimento di appartenenza ad una comunità, possa aiutare a migliorare le previsioni di alluvioni. In particolare, i modelli matematici possono beneficiare delle osservazioni aggiuntive fornite da neofiti di meteo.

Questa tesi ha dimostrato come le reti di sensori sociali a basso costo statici e dinamici possono complementare ed integrare reti di sensori tradizionali fisici statici, allo scopo di migliorare la precisione delle previsioni di piena. Questo studio può essere una potenziale applicazione nei recenti sforzi per costruire osservatori d'acqua in cui i cittadini non solo possono giocare un ruolo attivo nella raccolta, valutazione e comunicazione di informazioni idrologiche, ma può anche aiutare a migliorare i modelli matematici e quindi aumentare la resilienza dalle inondazioni

Maurizio Mazzoleni

Delft, Olanda

Contents

1

INTRODUCTION

1.1 BACKGROUND

Non-structural measures, such as Early Warning Systems (EWS), have been widely used in recent years in order to better predict floods and reduce their impact on urbanized areas. In most cases, hydrological and hydraulic models, with various degrees of complexity, are the main components of EWSs. Large amounts of data are needed to calibrate and validate such models, or update them in real-time. However, data are often limited, scarce or inadequate. On the other hand, new data sources are emerging, including citizen data, creating new opportunities and also new challenges.

1.1.1 Flood forecasting and early warning systems

In the last decades, the impact of floods has drastically increased worldwide (European Environment Agency, 2006; Di Baldassarre et al., 2010; Aerts et al., 2014; Dankers et al., 2014). Flood events in Europe, such as the 2002 Elbe floods and the 2007 UK floods, are considered national crises and are estimated to have caused around 15 and 6.5 billion Euro of damage respectively (European Environment Agency, 2006). Moreover, due to the combined effects of rapid urbanization, growth of population near to floodplains and flood levels increasing due to climate change and sea level rise, this trend seems likely to worsen in the near future (Hinkel et al., 2014; Jongman et al., 2014). In fact, societies seem to have the tendency to settle near water courses and this can be demonstrated by the fact that, according to the United Nations (2012), nine of the 10 largest urban agglomerates in the world are located in deltas or floodplain areas (Di Baldassarre et al., 2015). Despite the continuous construction of flood protection works, such as levees or reservoirs, flooding is still an important issue in many countries (European Environment Agency, 2006; Wilby et al., 2008). For this reason, the demand for tools to forecast water levels in rivers has significantly increased (Solomatine and Wagener, 2011).

Non-structural measures, such as flood EWS, allow for an accurate and timely real-time forecasting of river water levels and allow decision makers to take the most effective decisions in sufficient time to reduce the possibility of harm or loss (Todini et al., 2005; McLaughlin, 2002). Such measures are crucial for the proper evaluation of the flood risk and significantly reduce the direct and indirect costs of a flood in urbanized areas (Teisberg and Weiher, 2009). Among the various types of water system models, hydrological and hydrodynamic models are the most utilised in flood EWS in river basins. The inputs tosuch models are weather forecasting products, such as the rainfall forecast up to 5 days ahead provided by the European Centre for Medium-Range Weather Forecast (ECWMF). Bartholmes and Todini (2005) coupled a semi-distributed hydrological model (1km resolution) with several European meteorological models within the Po River Basin. Poor results are obtained for quantitative precipitation forecasting. In Todini et al. (2005) various examples of flood EWS implemented in different countries are reported.

In fact, deterministic predictions contain an intrinsic uncertainty due to the many sources of error which propagate through the model and therefore affect its output (Pappenberger et al., 2006). For this reason, EWSs should be able to quantify uncertainty around a given deterministic value. A growing number of studies analysed the impact of rainfall uncertainty on flood event prediction (Kavetski et al., 2006; Moulin et al., 2009; McMillan et al., 2011). One possible way to quantify uncertainty in flood forecasting is to use an ensemble of weather predictions. For example, ECWMF produces an ensemble of 51 different weather forecasts. The spread of such ensemble members is an indication of the uncertainty related to the meteorological model, the main source of uncertainty in flood forecasting systems (Buizza, 2008; Nester et al., 2012). However, in current operational flood forecasting systems, uncertainty is considered as the difference between observed and forecasted values.

Various studies have been carried out in order to assess uncertainty in flood forecasting systems (Todini et al., 2005; Xuan et al., 2009; van Andel et al., 2013; Plate and Shahzad, 2015). Some examples are reported below. In Liu et al. (2005) a formal Bayesian inference technique was applied to a distributed hydrological model to perform parameter sensitivity analysis and provide an estimate of predictive uncertainty in the Upper Xixian catchment (Huaihe River). Rao et al. (2011) developed a flood forecast model for the Godavari Basin in India by means of a semi-distributed modelling approach based on distributed deterministic inputs. Results showed good accuracy in flood peak estimation, with an increase in the lead time by 12 hours compared to conventional methods of forecasting. De Roo et al. (2003) developed a European-scale flood forecasting system (EFFS) using ECMWF's deterministic and ensemble prediction system as input for a distributed hydrological-hydraulic modelling framework. Thielen et al. (2009) presented the European Flood Alert System (EFAS) in order to provide local water authorities with medium-range (lead time from 3 to 10 days) probabilistic flood forecasting in order to increase preparedness and promote a culture of risk prevention. Jaun and Ahrens (2009) studied the advantages and limitations of probabilistic forecasting systems as opposed to deterministic ones.

1.1.2 Hydrological and hydrodynamic modelling

Hydrological models are tools used to represent a catchment's response to climate and/or land use variability in order to assess the flow hydrograph at the concentration point of the basin for purposes such as real-time flood forecasting (Solomatine and Wagener, 2011). Hydrological models can be divided into three distinct classes using the classification proposed by Wheater et al. (1993), namely mechanistic (i.e. physically-based models), parametric (known as grey box or conceptual models) and metric (called empirical, black box, or data-driven models). In Figure 1.1, a representation of these classes is given. Hydrological models can also be classified according to the spatial discretization of the model itself as distributed, semi-distributed or lumped (Solomatine and Wagener, 2011).

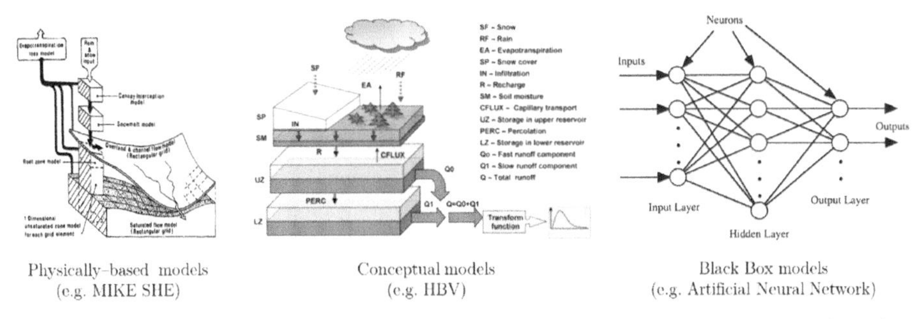

Figure 1.1. Hydrological models classification (adapted from Xevi et al. (1997); Dawson and Wilby (2001); Shrestha et al. (2009)

In physically-based models, the hydrological cycle including surface flow, unsaturated zone flow and groundwater flow is described by means of mathematical equations: the St. Venant, the Richards and the Boussinesq equations respectively. Usually this kind of model requires a significant amount of information in order to properly describe all the physical aspects of the phenomena. The distributed nature of this type of model allows for taking into account the spatial variability of the inputs and outputs in the catchment. However, spatial scale effects or simply lack of data may introduce significant uncertainties in the parameter values with consequent uncertainties in the predictions (Beven, 2001). For this reason, considering the limited data available in practice, the application of these types of models in operational real-time flood forecasting is quite difficult. Examples of physically-based model systems are MIKE SHE (DHI, 1998) and the Representative Elementary Watershed framework (REW, Reggiani et al., 1998).

Conceptual models describe the hydrological cycle in a river basin using the continuity equation. The structure of the model, i.e. how the physical processes are discretised in the basin, is specified a-priori by the modeller and it is not derived from rainfall-runoff data observation. Instead, it comes from an understanding of the hydrological response. Conceptual models can be very useful in operational practice. A typical example of a conceptual model is given by the rainfall-runoff model HBV (Hydrologiska Byras Vattenbalansavdelning, Lindström et al., 1997) which consists of four main modules, namely 1) snowmelt and snow accumulation, 2) soil moisture and effective precipitation module, 3) evapotranspiration and 4) runoff response. Input data in HBV are observations of precipitation, air temperature and estimates of potential evapotranspiration. Other examples of conceptual models are the GR4J model (Perrin et al., 2003), NAM-MIKE11 (Havnø et al., 1995; Shamsudin and Hashim, 2002), PCR-GLOBWB model (van Beek et al., 2011), Sacramento model (Burnash, 1995), and the Probability Distributed model (Moore, 1985, 2007), among others.

Black box models (BBMs) are based on relationship(s) that best connect inputs with observations without taking into account the mathematical formulations of the physical processes involved. An early example of this type of model is the so-called Instantaneous Unit Hydrograph (IUH), which has been actively used for dozens of years. During the last decades, statistical and data driven models (DDMs) have been introduced to describe various types of relations between inputs and outputs. These models use regression equations to estimate the flow hydrograph (or another type of output) from past observations of discharge or other physical variables. The simplest DDM is the linear regression model. Nowadays there are a host of nonlinear and sophisticated DDMs such as auto-regressive models, artificial neural networks (ANN, Tokar and Johnson, 1999; Dawson and Wilby, 2001; Dibike and Solomatine, 2001; Govindaraju and Rao, 2013), fuzzy rule-based systems (Bardossy et al., 1995), fuzzy regression (Bardossy et al., 1990; Kim et al., 1996; Özelkan and Duckstein, 2000), genetic programming (Whigham and Crapper, 2001; Rabuñal et al., 2007), and support vector machines (Dibike et al., 2001), etc.

Hydrodynamic models are useful tools to estimate water level, flow velocity and flood extent in rivers and their flood-prone areas. These types of models produce a numerical solution of the continuity and momentum equations in case of unsteady flow adopting either finite difference, finite element or finite volume methods to solve the Saint Venant and the Navier-Stokes equations. They can be divided into one-dimensional 1D models (e.g. HEC-RAS, MIKE11 and ISIS), two-dimensional 2D models (as LISFLOOD-FP, SOBEK, Mike21, InfoWorks2D, TELEMAC-2D etc.) and three-dimensional 3D models (e.g. CFX, FLUENT and PHEONIX). The latter are rarely used, but one can observe an increasing number of applications of 2D modelling by environmental agencies and water authorities.

Despite the advantages of 2D models, 1D hydrodynamic models are still widely employed by practitioners due to their capability to describe the hydraulic behaviour of many natural rivers (e.g., Pappenberger et al., 2006). Usually, 1D models solve the De Saint Venant equations through a numerical scheme, such as the four point implicit approach (box scheme, Preissmann, 1961), to discretize the continuity and momentum equation in case of unsteady flow in open channel. As it can be seen in Figure 1.2, the river behaviour is represented using river cross-sections (shown in black) perpendicular to the main channel (blue curve). The cross-sections can include both the main channel and flood-plain area, as an attempt to capture the flood wave propagation over the flood plain in onedimension. The simplified approach used by 1D models to solve the problem of the interaction between channel and floodplain, is to divide the system into a number of separate channels and write continuity and momentum equations for each one, assuming a horizontal water surface at each cross section normal to the direction of flow. However, in the case of low return period floods and consequent low water level, it might be difficult to represent the flood propagation along the floodplain using a 1D model. For this reason, it might be more accurate to use a 1D model only inside the main channel and a 2D model on the flood-plain area.

2D hydrodynamic models have proved to be useful tools in simulating river hydraulics and floodplain and flood-prone area inundation processes under uncertainty for flood risk management (Horritt, 2006). Commonly, 2D models are based on the numerical solution of the two dimensional Navier-Stokes equations for an incompressible fluid with constant density.

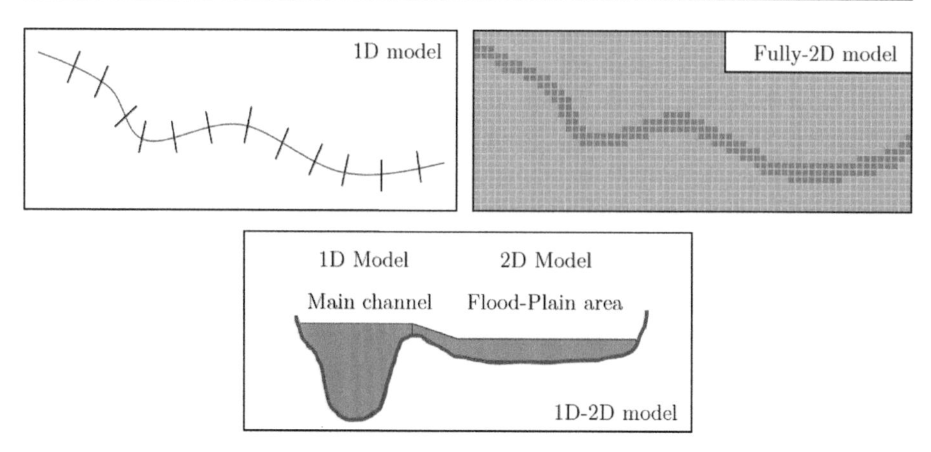

Figure 1.2. Hydrodynamic models classification (Ranzi et al., 2011)

In 2D models the main channel and the flood-prone area are represented by the Digital Elevation Model (DEM) of the area (see Figure 1.2). Fully-2D numerical modelling usually provides more accurate results, in terms of inundated area, but demands a significant effort, particularly when high spatial resolution is required. One of the main drawbacks of a fully-2D model approach is the heavy computational requirement for large study areas. One possible solution to reduce such costs is the implementation of simplified 2D code like the one proposed by Bates and De Roo (2000). Another possible solution is to use 1D-2D coupled models (see Figure 1.2). These models are frequently applied in large floodplains or polders with complex topography. In this way it is possible to study the one-dimensional behaviour of the main channel with the 1D model and then to interpret the propagation of the flood wave in the floodplain by the 2D model, avoiding the onerous description of the whole riverbed geometry in 2D and, consequently, achieving a reduction in the computational time (Aureli et al., 2006).

In order to assess hydrological and hydraulic model performance compared to observed values, different statistical measures have been proposed in the past years. However, in this thesis, only the three of them are used, described as follows. One of the most widely used measures in hydrology is the Nash-Sutcliffe Efficiency (NSE) index (Nash and Sutcliffe, 1970) which compares simulated and observed time series as:

$$NSE = 1 - \frac{\sum_{t=1}^{T}(WD_t^m - WD_t^o)^2}{\sum_{t=1}^{T}(WD_t^m - \overline{WD_t^o})^2} \quad (1.1)$$

where WD_t^m is the simulated water depth in the t-th time step, WD_t^o is the observed water depth, $\overline{WD_t}^o$ is the average observed water depth while T is the number of pairs of simulated and observed water depths. A NSE equal to 1 represents a perfect model simulation while, a NSE equal to zero indicates that the simulated streamflow is as accurate as the mean of observed water depth.

The Pearson Correlation coefficient (R) is used to measure the linear correlation between two variables, in this chapter simulated and observed water depth.

$$R = \frac{COV(WD^m, WD^o)}{\sigma(WD^m)\sigma(WD^o)} \tag{1.2}$$

where $\sigma(WD^m)$ and $\sigma(WD^o)$ are the standard deviation of simulated and observed water depth, respectively. Values of R close to 1 indicate a positive correlation between the two variables. The last index, bias index (bias), measures the tendency of the simulated water depth to be larger or smaller than the observed water depth.

$$Bias = \frac{\sum_{t=1}^{T} WD_t^m}{\sum_{t=1}^{T} WD_t^o} \tag{1.3}$$

Values greater than 1 indicate overestimation of the water depth while values smaller than 1 represents an overall underestimation.

1.1.3 Uncertainty in hydrological and hydrodynamic modelling

The reliable characterization and reduction of the uncertainties affecting the modelling of hydrological and hydrodynamic processes is an important scientific and operational challenge (Wagener and Gupta, 2005; Quinonero-Candela et al., 2006; Liu and Gupta, 2007; Renard et al., 2010). In fact, uncertainty estimation is crucial for risk-based decision making processes in water resources management (Pappenberger and Beven, 2006) as it adds 'honesty' and reliability to model output (Krzysztofowicz, 2001). Uncertainty can be due to either the inherent stochastic nature and variability of hydrological processes, i.e. aleatory uncertainty (Koutsoyiannis, 2010; Montanari and Koutsoyiannis, 2012), or to our imperfect state of knowledge of the hydrological system and our limited ability to model it, i.e. epistemic uncertainty (Merz and Thieken, 2005; Hall and Solomatine, 2008; Domeneghetti et al., 2013). Four main sources of uncertainty can be identified in the hydrological and hydraulic models used in flood forecasting systems (Pappenberger et al., 2006; Liu and Gupta, 2007; Götzinger and Bárdossy, 2008; Solomatine and Wagener, 2011):

1. Input uncertainty: approximation of the observed hydrological variables used as input or calibration data (e.g. rainfall, temperature and river discharge);

2. Output uncertainty: for example, rating curve errors affecting runoff estimates or unreliable flow hydrograph estimation;
3. Parametric uncertainty: induced by imperfect model calibration due to uncertain calibration data, model approximations, imperfect process understanding, etc.;
4. Structural uncertainty: induced from assumptions, simplifications and approximations made to conceptualise complex hydrological processes, made considering study purpose, cost, computational resources and time.

Other authors have also quantified uncertainty in flood plain modelling related to the uncertain operation of hydraulic structures connecting rivers and wetlands (Alfonso and Tefferi, 2015).

Since each model is a mathematical schematization of some natural physical process, a proper definition of the model structure and parameters is necessary in order to correctly represent the behaviour of the catchment and reduce the model uncertainty (Pappenberger et al., 2006; Di Baldassarre and Montanari, 2009). In addition, even in the case of perfect model structure and parameter estimation, an uncertain and inadequate characterization of rainfall inputs can cause imprecise runoff predictions (Beven, 2001). Several research activities aimed at reducing such uncertainty (Raiffa and Schlaifer, 1961) in streamflow and water level estimation have been carried out due to its importance in deciding whether to issue a flood warning.

Different approaches have been proposed for quantifying the uncertainty in hydrologic predictions (Shrestha and Solomatine, 2008). In Monte Carlo approaches, the quantification of model output uncertainty, resulting from uncertain model parameters, input data or model structure, is achieved by random sampling from the distribution of uncertain input. One of the main drawbacks of Monte Carlo approaches is the large number of samples required (Shrestha and Solomatine, 2008). Generalized Likelihood Uncertainty Estimation (GLUE, Beven and Binley, 1992) is one of the most popular versions of Monte Carlo approaches for uncertainty analysis in hydrology. Critical analyses of the GLUE approach are presented in Mantovan and Todini (2006) and Stedinger et al. (2008). Different Bayesian approaches like standard Bayesian approaches (Kuczera and Parent, 1998; Krzysztofowicz, 1999; Feyen et al., 2007), Bayesian Recursive Estimation (Thiemann et al., 2001), Bayesian hierarchical models (Kuczera et al., 2006; Kavetski et al., 2006; Huard and Mailhot, 2008) and Bayesian model averaging (Duan et al., 2007; Marshall et al., 2007) are also used in uncertainty analysis. The last decade has witnessed a major shift from deterministic to probabilistic flood hazard assessment by means of hydrodynamic modelling. Applications of uncertainty analysis in hydraulic modelling are reported in Pappenberger et al. (2006), Di Baldassarre et al. (2009), Moel and Aerts (2010), Brandimarte and Baldassarre (2012), Brandimarte and Woldeyes (2013), Domeneghetti et al. (2013), Yan et al. (2013), Mazzoleni et al. (2014, 2015), Mukolwe et al. (2015) and Alfonso et al. (2016).

1.1.4 Data assimilation

Over the past decades, model updating techniques, and particularly data assimilation approaches, have been used within water system models (WMO, 1992; Refsgaard, 1997) for reducing predictive uncertainty. Data assimilation methods are widely used in water-related applications to optimally update model states, inputs or parameters as a response to real time observations of hydrological variables, usually measured by physical sensors (Heemink and Kloosterhuis, 1990; Robinson et al., 1998; McLaughlin, 2002; Moradkhani et al., 2005b; Walker and Houser, 2005; Liu and Gupta, 2007; Reichle, 2008).

Among data assimilation techniques the Kalman filter is one of the most used methods to assimilate, in an efficient recursive way, observed noisy data into dynamic systems (Kalman, 1960). However, one limitation of the Kalman filter is that it is optimal only for linear dynamic systems. For this reason, different variants of the Kalman filter, such as the extended Kalman filter (Madsen and Cañizares, 1999; Aubert et al., 2003), unscented Kalman filter, ensemble Kalman filter (Reichle, 2000; Evensen, 2003; Komma et al., 2008; Mendoza et al., 2012; Noh et al., 2013; Rafieeinasab et al., 2014) and recursive ensemble Kalman filter (McMillan et al., 2013) have been proposed and applied in hydrologic modelling. Madsen and Cañizares (1999) compared the performance of EKF and EnKF in coastal area modelling. Although the study showed that the EnKF does not fail in the case of strong non-linear dynamics, it was found to be very time consuming. Application of Kalman filtering methods to hydrodynamic modelling has been explored by Verlaan and Heemink (1995) and Verlaan (1998).

Yet another version of a non-linear filter is the particle filter (PF) (Arulampalam et al., 2002), which has also been used in flood forecasting tasks (Moradkhani et al., 2005b; Noh et al., 2014; Salamon and Feyen, 2009). In PF, the posterior density function is represented by a set of random samples with associated weights according to the full prior density and resampling approach used (see e.g. Arulampalam et al., 2002; Weerts and El Serafy, 2006). The computational requirements (much higher than those of the Kalman filters) and problems with nearly noise-free models are seen as the main disadvantages of the PF.

In contrast to the previous sequential methods, variational assimilation methods have been widely used in weather forecasting and costal engineering applications (Li and Navon, 2001; Seo et al., 2003; Valstar et al., 2004; Fischer et al., 2005; Lorenc and Rawlins, 2005; Seo et al., 2009; Lee et al., 2011a, 2012; Liu et al., 2012). In these methods, the cost function that measures the difference between the error in the initial conditions and the error between model predictions and observations over time is minimised to identify the best estimate of the initial state condition (Seo et al., 2009; Lee et al., 2011a). A detailed review of the status, progress, challenges and opportunities in advancing DA for operational hydrologic predictions is provided in Liu et al. (2012).

The mentioned DA methods require a significant amount of real-time data in order to update hydrological and hydrodynamic models and improve flood forecasts. In fact, due to the complex nature of hydrological processes, spatially and temporally distributed measurements are needed in the model updating procedures to ensure a proper flood prediction (Clark et al., 2008; Rakovec et al., 2012; Mazzoleni et al., 2015a). Hydrological observations, used to update the water model states, can include streamflow (Pauwels and De Lannoy, 2006; Weerts and El Serafy, 2006; Pauwels and De Lannoy, 2009), snow cover (Andreadis and Lettenmaier, 2006), soil moisture (Brocca et al., 2010, 2012) or water level observations coming from in situ sensors (Madsen and Skotner, 2005; Neal et al., 2007) and remote sensing (Giustarini et al., 2011). Aubert et al. (2003) integrated distributed values of soil moisture and streamflow from physical sensors into a lumped conceptual hydrological model by means of an extended Kalman filter. Streamflow prediction is improved by the assimilation of both soil moisture and streamflow individually and by coupled assimilation. Cao et al. (2006) proposed a calibration approach based on integration of multiple internal variables with multi-site locations, resulting in a more realistic parameterization of the hydrological process. De Lannoy et al. (2007) assimilated distributed values of soil moisture in an agricultural field, assessing the influence of the biased or the bias-corrected state estimates into a biased model. They pointed out that the results are dependent on the nature of the model itself. In fact, in the case of a model that is only biased for soil moisture it is better to post-process the soil moisture with the bias analysis than update the model states since the large increment of updated soil moisture might result in an incorrect water balance. Lee et al. (2011) assimilated streamflow and in situ soil moisture observations into a distributed hydrological model showing that the integration of streamflow observations at interior locations, in addition to those at the outlet, improves soil moisture and streamflow prediction along the channel network. Mendoza et al. (2012) evaluated the performance of a flood forecasting scheme assimilating sparse streamflow observations using an Ensemble

Kalman Filter. They found that, for the considered case study, the hydrologic process representation for the upper part of the basin is the major source of uncertainty. In Xie and Zhang (2010), Rakovec et al. (2012), Lee et al. (2012) and Chen et al. (2012) the distributed streamflow observations were assimilated in hydrological models with different structures. Overall, the authors found that assimilation of observations from inner points of the basin helps to further improve the hydrograph estimation. Moreover, they demonstrated that assimilation performances are more sensitive to the spatial distribution of sensors rather than to the updating frequency.

Many of the previous studies are related only to in-situ observations (discharge, soil moisture, etc.). However, the increasing availability of distributed remote sensing data has stimulated different research activities in order to assimilate these data into hydrological and hydrodynamic models. Examples of assimilation of surface water elevation into hydrodynamic models can be found in Montanari et al. (2008), Neal et al. (2009) and Giustarini et al. (2011). In addition, physical variables such as soil moisture and snow cover area can be assimilated in order to increase the reliability of water models (Brocca et al., 2010; Reichle et al., 2008; Brocca et al., 2012). All these methodologies demonstrated an increase in the model performance after the assimilation of the remote sensing data. On the other hand, the assimilation of remote sensing data in an EWS is not an easy task due to the temporal availability of the remote information, which might not correspond to the occurrence of the flood event.

Most hydrological and hydrodynamic models are calibrated for given hydrometeorological conditions, but due to the fact that weather conditions might change it is necessary to assume changing model behaviour in time. For this reason it is crucial to consider the adjustment of model parameters together with state variables over time (Moradkhani et al., 2005a; Montaldo et al., 2007). The idea is to have change the model parameters as new observations are assimilated. An interesting example of an interactive dual state-parameter estimation (both model states and parameters will be simultaneously estimated) by means of a Kalman filter in the context of hydrology is proposed by Todini et al., (1976). Moradkhani et al. (2005a) proposed an integrated framework for dual state-parameter estimation using EnKF leading to ensemble streamflow forecasting. In addition, the authors proposed a recursive algorithm which does not require storage of all past information, as in case of batch calibration. The dual EnKF uses the ensemble of model trajectories in an interactive parameter-state space and provides the confidence interval of parameter-state estimation. Lü et al. (2011) developed a method to couple optimal parameter estimation and EKF in order to estimate root zone soil moisture. Such study assumes that model parameters do not change randomly in time but they do have certain unknown values. Chen et al. (2008) showed that the smoothing ensemble Kalman filter, used to simultaneously estimate model states and parameters (of an ecosystem model) by concatenating unknown parameters and state variable into a joint state vector, improves flux estimation and reduces parameter uncertainty.

It should be noted that in operational practice it is preferred to correct the model inputs (in most cases), states, initial conditions and parameters in an empirical and subjective way rather than by applying advanced data assimilation techniques for improving hydrologic forecast (Seo et al., 2009). It is in fact assumed that uncertainties in the input data are the main source of uncertainty in operational flood forecasting (Sittner and Krouse, 1979; Canizares et al., 1998; Todini et al., 2005). Liu et al. (2012) pointed out that the need for implementing reliable data assimilation methods in operational forecast is increasing in order to fill this gap between the scientific world and practice.

Besides water-related applications, model updating techniques are frequently used in other fields due to necessity to assimilate and integrate new measurements coming from various locations and at various times. Below, a brief review of some applications of data assimilation in non water-related research fields is presented. Assimilation of new observations in control systems and robotics is a topical problem which has been addressed by various authors. For example, Moulton et al. (2001) proposed a fuzzy error correction control system to navigate a robot along a modifiable path. Fong (2009) developed a multi model adaptive filter for use in a multi-sensor track fusion system for target tracking. Hover (2009) proposed the solution of a mobile planning sensor problem combining an extended Kalman filter and a classical optimization scheme in order to minimise the error in the model. This method allows for analysing the complex trajectory and vehicle dynamics using data collected for near real-time assimilation. The problem of tracking a tactical ballistic missile, in the area of air defence, is complicated due to the variability in the boost, exo-atmospheric and endo-atmospheric phases of flight. The idea proposed by Cooperman (2002) is to develop a tactical ballistic missile tracker within an Interacting Multiple Model (IMM) framework which simultaneously weights all model states and then adapts them to the one most closely matching the data based upon measurement residuals. From this paper it is concluded that the tracking accuracy improved compared to the use of a single sensor. In Shima et al. (2002) an efficient multiple model adaptive estimation (MMAE) in ballistic missile interception is presented. In neurology, data assimilation it is used for assessing brain deformation and tumour growth based on mathematical models and medical images of different patients (Lunn et al., 2006; Ji et al., 2009; McDaniel et al., 2013). The idea is to forecast the future evolution of the tumour and its composition in order to find the most efficient medical care. The results of McDaniel et al. (2013) demonstrate the use of ensemble forecasting and data assimilation to make improved estimations the of future growth of a simulated glioblastoma given synthetically generated observations of the tumour.

1.1.5 Citizen Science

As previously described, the hydrological sciences are important to correctly predict floods and reduce loss of life. Traditionally, static physical sensors, such as pressure sensors, water level sensors, heat flux sensors and pluviometers, are used by water authorities in hydrological and hydrodynamic models. These data are used to calibrate, validate or update such models in real-time, to use them in EWSs for water depth and flow prediction. However, a main problem, highlighted by Hannah et al. (2011) (among others), is the scarcity of data in both spatial and temporal domains. Such problems can be related to the fact that traditional physical sensors require proper maintenance and personnel which can be very expensive in case of a large network.

Over the last couple of decades technological improvements have led to the spread of heterogeneous networks of low-cost sensors used to measure hydrological variables, such as water level or precipitation, in a more distributed way (Yarvis et al., 2005). The main advance of using these type of sensors is that they can be used not only by technicians, as for observations from traditional physical sensors, but also by regular citizens, and that due to their reduced cost, a more spatially distributed coverage can be achieved. Recently, citizen science activities have been widely promoted in order to collect crowdsourced (CS) observations of hydrological variables ,generate additional knowledge of the water cycle, and use such knowledge in decision making (Bonney et al., 2014; Buytaert et al., 2014).

Howe (2008) defined the concept of crowdsourcing as "the act of taking a job traditionally performed by a designated agent (usually an employee) and outsource it to an undefined, generally large group of people in the form of an open call". However, a main problem in citizen science is the motivation that drives citizens to be involved in such activities. In addition, Bonney et al. (2014) stated that "Despite the wealth of information emerging from citizen science projects, the practice is not universally accepted as a valid method of scientific investigation. Scientific papers presenting volunteer-collected data sometimes have trouble getting reviewed and are often placed in outreach sections of journals or education tracks of scientific meetings. At the same time, opportunities to use citizen science to achieve positive outcomes for science and society are going unrealized." Buytaert et al. (2014) pointed out that motivations of citizen engagement vary according to geographical location. In fact, citizen science projects in wealthy regions aim to increase awareness and scientific literacy, while in developing regions the main goals are more related to enhancement of community well-being such poverty alleviation (Gura, 2013). Once the citizens are involved, different levels of engagement can be found. In fact, engagement can be driven either by egoism (increasing one's own welfare), collectivism, (increasing the welfare of a specific group that one belongs to), altruism (increasing the welfare of another individual or group of individuals), principalism (upholding one or more principles dear to one's heart), or to follow a moral principle (Batson et al., 2002). However, the engagement is dynamic and may evolve during the citizen's involvement period. For example, what drives someone one day to volunteer may not keep doing it tomorrow (Rotman et al., 2012). Bonney et al. (2009) proposed three different approaches, defined as contributory, collaborative and co-created.

Recently, Gharesifard and Wehn (2016) studied the drivers and barriers for sharing citizen-sensed weather data via online amateur weather networks. A detailed and interesting review of the examples of citizen science applications in hydrology and water resources science is reported by Buytaert et al. (2014). In this study it is pointed out that, in most cases, the final scope of citizen activities is connected to water quality. In fact, in the case of hydrological applications, streamflow measurements are complex by nature and difficult to directly infer. A possible solution could be the use of camera-based water level measurements (Royem et al., 2012). An example of low-cost sensor used to measure water level can be a staff gauge (the reference sensor) connected with a Quick Response (QR) codes used to infer the spatial location of the measurement. The idea is that, instead of a complex installation of a static physical sensor with all the required components, citizens equipped with a mobile phone (the dynamic sensor), would take measurements assessing the water level at the staff gauge location. In addition, data transmission is improving due to increasing mobile phone penetration around the world.

In hydrological applications, various projects have been initiated in order to assess the usefulness of CS observations inferred from low-cost sensors owned by citizens. For instance, in the project CrowdHydrology (Lowry and Fienen, 2013), a method to monitor stream stage at designated gauging staffs using crowd sourced text messages of water levels is developed using untrained observers. Cifelli et al. (2005) described a community-based network of volunteers (CoCoRaHS), engaged in collecting precipitation measurements of rain, hail and snow. An example of hydrological monitoring of rainfall and streamflow values, established in 2009 within the Andean ecosystems of Piura, Peru, based on citizen observations is reported in Célleri et al. (2009). Degrossi et al. (2013) used a network of wireless sensors in order to map the water level in two rivers passing by Sao Carlos, Brazil. iSPUW Project aims to integrate data from advanced weather radar systems, innovative wireless sensors and crowdsourcing of data via mobile applications in order to better predict flood events in the urban water systems of the Dallas-Fort Worth Metroplex (Seo et al., 2014; ISPUW, 2015). Other examples of crowdsourcing water-related information include the Crowdmap platform for collecting and communicating information about the floods in Australia in 2011 (ABC, 2011), and informing citizens about the proper time to drink water in an intermittent water system (Au et al., 2000; Alfonso, 2006; Roy et al., 2012). Air quality observations, provided by volunteers by means of mobile technologies, have been used in the CITI-SENSE project (Castell et al., 2015; Schneider et al., 2015). Recently, Cortes Arevalo (2016) proposed a method to include volunteers' information to support proactive inspection of hydraulic structures.

A drawback of using CS observations is related to the intrinsic low variable accuracy, due to the lack of confidence in the data, and the variable life-span of each individual sensor with the consequent intermittent nature of the observations. For this reason, an important aspect in assimilating CS observations in hydrological and hydrodynamic modelling is the correct evaluation of the observation accuracy. According to Bordogna et al. (2014) and Tulloch and Szabo (2012), quality control mechanisms should consider contextual conditions to deduce indicators about reliability (expertise level), credibility (volunteer group) and performance of volunteers such as accuracy, completeness and precision level. Bird et al. (2014) addressed the issue of data quality in conservation ecology by means of new statistical tools to assess random error and bias in such observations. Cortes Arevalo et al. (2014) evaluated data quality by distinguishing the in-situ data collected from a volunteer and from a technician and comparing the most frequent value reported at a given location. They also gave some ranges of precision according to the rating scales. With in-situ exercises, it might be possible to have an indication of the reliability of data collected (expertise level). However, this indication does not necessarily lead to a conclusion about the degree of accuracy of the observer and it might not be enough at operational level to define accuracy in data quality. In fact, every time a crowdsourced observation is received in real-time, itsreliability and accuracy should be identified. To do so, one possible approach could be to filter the measurements following a geographic approach which defines semantic rules governing what can occur at a given location (e.g. Vandecasteele and Devillers, 2013). Another approach could be to compare measurements collected within a pre-defined time-window in order to calculate the most frequent value, the mean and the standard deviation. In this way it can be possible to assess data quality for a given citizen in real-time.

As described in the previous section, DA applications require specific, frequent and high quality measurements, which may not be compatible with the distributed, intermittent and, potentially, lower-quality nature of citizen-based data (Shanley et al., 2013; Buytaert et al., 2014; Lahoz and Schneider, 2014). That is why interpolation and merging techniques are commonly used to integrate citizen observations within mathematical models.

Kovitz and Christakos (2004) assimilated fuzzy data sets assigning probabilities of plausible events based on general knowledge through information maximization and then applying a Bayesian maximum entropy method. Schneider et al. (2015) reported an example of data fusion used to provide a combined concentration field by regressing dynamic air quality observations against model data and spatially interpolating the residuals (see Figure 1.3).

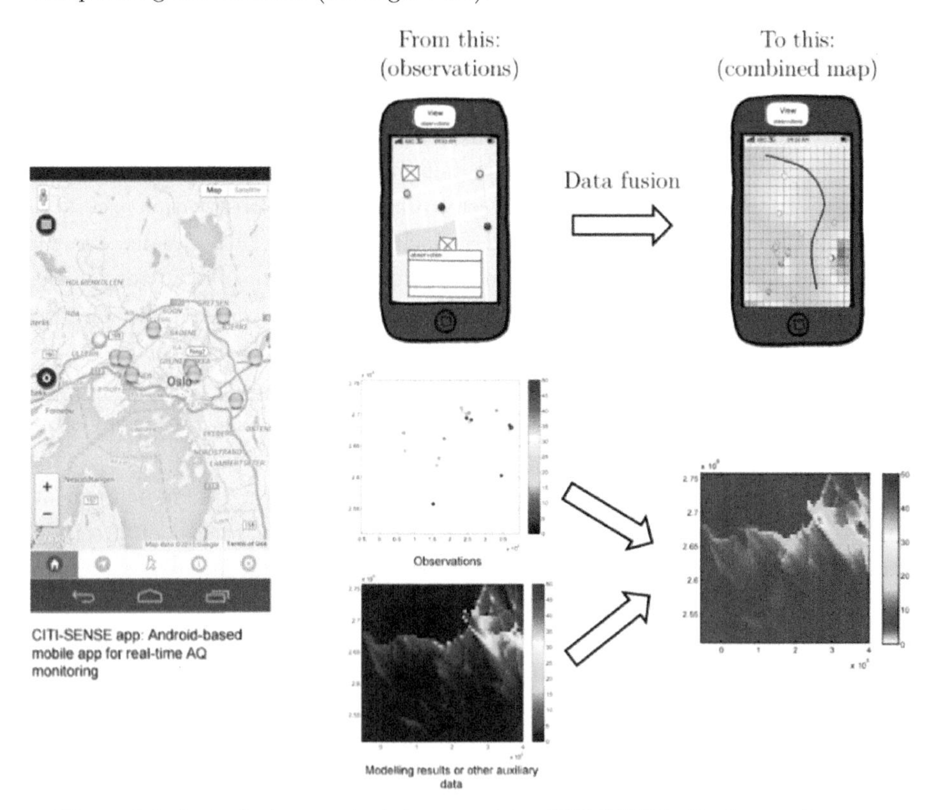

Figure 1.3. Data fusion methodology used in CITI-SENSE Project to integrate dynamic air quality observations (adapted from Schneider et al., 2015)

Beside DA applications, intermittent (or short duration) and distributed data can be used for model calibration as shown by Sheffield et al. (2006) and Seibert and Beven (2009). Aronica et al. (1998) proposed a fuzzy-rule-based calibration to compare model predictions and highly uncertain information about the flood arising from several different different types of observations. Seibert and McDonnell (2002) proposed an approach to calibrate hydrological models using both hard and soft data (e.g. percent of new water, reservoir volume, etc.) provided by experimentalists. Vaché et al. (2004) demonstrated the effectiveness of using soft-data for multi-objective calibration of hydrological models.

1.2 Motivation

Current flood forecasting applications limit the use of CS observations. Although some efforts to validate model results against such observations have been made, these are mainly done in a post-event analysis (Aronica et al., 1998; Seibert and McDonnell 2002; Vaché et al. 2004; Sheffield et al. 2006; Seibert and Beven 2009). The added value of information coming from citizens, therefore, is not typically integrated into hydrological and/or hydraulic models

Nowadays, model updating occurs only in the form of data assimilation using measurements of streamflow, soil moisture, etc. coming from static physical stations. In fact, only a few recent studies considered the integration of crowdsourced observations with water-related models (Buytaert et al., 2014). Some applications, as mentioned before, are related to data fusion appliedto air quality sensors (Schneider et al., 2015). For this reason, the main motivation of this thesis is to fill the gap in hydrological applications of CS observations and investigate the effects of their assimilation, derived using different types of sensors, in hydrological and/or hydraulic models by developing an optimal model updating algorithm. The social motivation behind this research it is to bring citizens closer to decision making processes.

It is worth noting that the research presented in this thesis is carried out in the framework of the WeSenseIt (WSI) Citizen Observatory of Water Project (grant agreement no: 308429). The WSI project is funded by the Seventh Framework Programme for Research and Technological Development (FP7) of the European Union for a total duration of 48 months from October 2012 to October 2016. The WSI consortium is composed of 14 partners from 6 countries, including research and development organizations (Small Medium Enterprises, SMEs) that complement each other in terms of scientific and technical expertise.

Three different case studies, Doncaster (UK), Bacchiglione (Italy) and Delfland (The Netherlands) are considered within the WSI Project.

The main goal of the WSI project is to allow citizens and communities to become active stakeholders capturing additional information to be communicated to the water authorities to improve water model performance and, consequently, flood prediction, with a shift from the traditional one-way communication paradigm. The WSI concept is presented in Figure 1.4. In order to achieve its objectives, the WSI project developed and tested new technology for collecting physical and social CS data by means of innovative static and low-cost mobile sensors. Temperature, soil moisture, precipitation, water level, flow velocity, heat flux and snow depth sensors are the main ones developed within the WSI project. Such sensors are optimally integrated into a heterogeneous network of static and dynamic sensors optimally designed by means of new multi-objective optimisation methods.

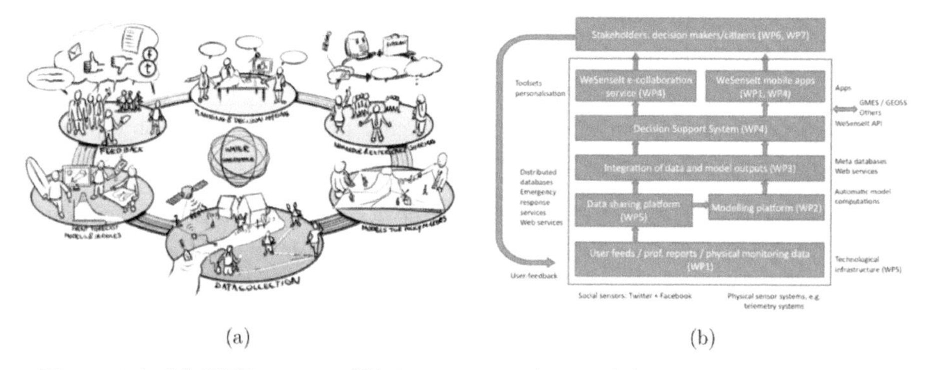

(a) (b)

Figure 1.4. (a) WSI concept (WeSenseIt, 2016), and (b) description of the Work Packages (WP) within the WSI project (WeSenseIt, 2016)

An example of strategic sensor locations in the Bacchiglione case study is shown in Figure 1.5.

The observations derived from the network of sensors are used to optimally assimilate observations coming from the heterogeneous network of sensors within hydrological and hydrodynamic models. That is why the present research contributes to the WSI project. An active citizen observatory is established by designing and implementing an e-collaboration environment for participation, feedback and decision-making. In particular, a mobile phone app where citizens can send flood reports and precipitation and water level sensor readings are developed within the WSI project (see Figure 1.6). Such readings are then sent to the WSI platform to be used by the case study partners in their hydrological and hydraulic models.

Finally, changes in the governance processes from the implementation of the innovative citizen observatory are analysed.

Figure 1.5. Sensor locations within the Bacchiglione case study (Italy)

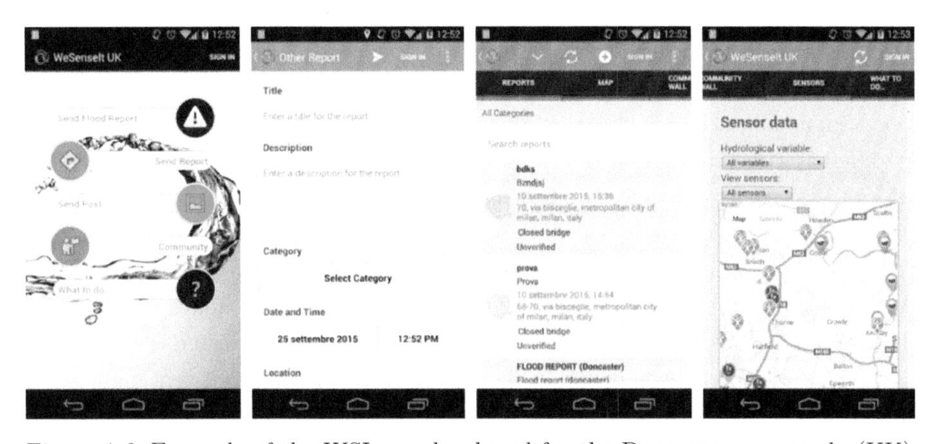

Figure 1.6. Example of the WSI app developed for the Doncaster case study (UK)

1.3 TERMINOLOGY

In this section, the definitions of some important terms which will be used in this PhD thesis are presented.

Sensor: a device that responds to a physical stimulus (as heat, light, sound, pressure, magnetism, or a particular motion) and transmits a resulting impulse, as for measurement or operating a control (following the Merriam-Webster online dictionary). Examples of sensors are precipitation gauges, level gauges, flow gauges, flow velocity gauges, etc.

Static Physical sensor (StPh): a sensor providing information at a given spatial location. Such sensors are characterized by having a known uncertainty, which does not change in time. An example of a static sensor is a pluviometer or a staff gauged used to measured water depth in a given location in a basin. The StPh sensors considered in this thesis are classic water level sensors (e.g. ultrasonic sensor) having low uncertainty (see Figure 1.7)

Dynamic sensor: device that provides information in different spatial locations. A typical example of a dynamic sensor is a smartphone used by a citizen to take a picture and infer the water levels at a particular location. These sensors send information without following a predefined spatial path. (It is worth noting that this thesis does not aim at developing methods to process such pictures in order to derive the numerical value of water depth but to use such results into water models.).

Social sensor: any data collection activity that is carried out by a citizen using either a sensor or communicating an observation by means of a mobile device like a smartphone (Ciravegna et al., 2013). These sensors can be static or dynamic. In case of dynamic social sensors, the observations will arrive from different locations at different time steps. For this reason, the main characteristic of this type of sensors is spatial randomness and variable uncertainty in time and space. Social sensors can be static (Static Social, StSc) or dynamic (Dynamic Social, DySc). The StSc sensors are staff gauges located in strategic point of the river, as shown in the WSI project description (see Figure 1.5), used by citizens to estimate water depth values. The WSI mobile phone app is used to send observations using the QR code as a geographical reference point. In the case of DySc sensors, a citizen might send the information related to the distance between the water profile and the river bank using the same WSI app of StSc at random locations along the river. It might be in fact difficult to estimate the water depth without having any indication about river depth. Knowing the river bed and bank elevations, it can be possible to estimate the water depth.

In this case, the observations have higher degree of uncertainty due to the indirect method used to estimate water depth. The main characteristics of CS observations provided by StSc and DySc sensors is the random accuracy and the variable spatio-temporal coverage due to the citizens' expertise and behaviour.

A- STATIC PHYSICAL SENSOR (StPh)

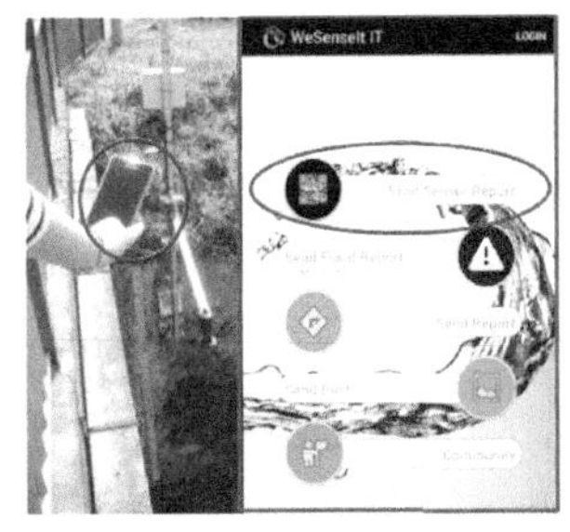

B - STATIC SOCIAL SENSOR (StSc)

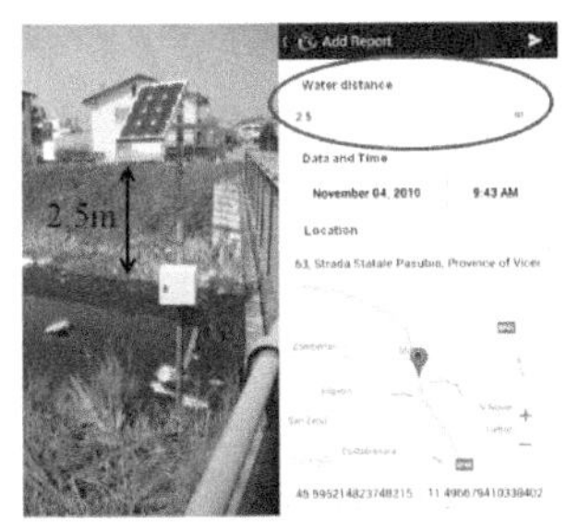

C- DYNAMIC SOCIAL SENSOR (DySc)

Figure 1.7. Sensor classification used in this thesis

Heterogeneous network of sensors: Two or more sensors of different types (e.g. static and social sensors) forming large, interconnected, and heterogeneous networks.

Crowdsourced observations: observations derived from a multiple source of sensors, with variable uncertainty and life-span (Howe, 2008). Examplesof crowdsourced observations are those provided by different citizens using smartphones to read the water level indication from a social sensor like a staff gauge. It is worth noting that in this thesis the arrival moment of the crowdsourced observation and the assimilation moment are considered as coincident.

Synchronous observations: observations that have the same sampling frequency as the model time step. Such observations can be regular (coming with the fixed frequency), or intermittent (variable frequency) (see Figure 1.8).

Asynchronous observations: observations received by the model at a higher frequency than the model time step. For example, in case of model time step of 1 hour, asynchronous observations might be received at any moment within this hour (see Figure 1.8).

Regular observations: situation in which at least one observation is received at each model time step. Examples of regular observations are the ones provided by physical sensors (see Figure 1.8).

Intermittent information: social sensor providing observations with different life-spans. In fact, the information from a specific sensor might be sent just only once, occasionally, or in time steps that are non-consecutive (see Figure 1.8).

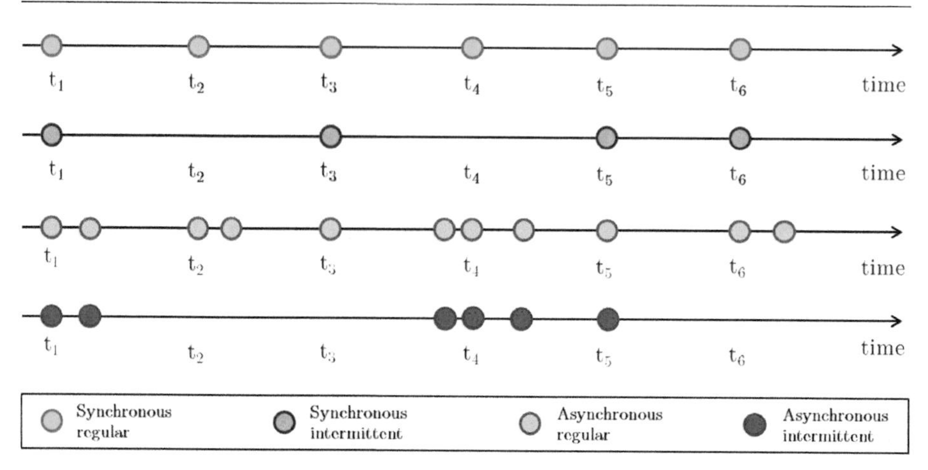

Figure 1.8. Representation of synchronous, asynchronous, regular and intermittent observations, where time is the model time step

1.4 RESEARCH OBJECTIVES

The main objective of this thesis is to investigate methods of assimilating crowdsourced observations from social and physical sensors with varying spatial and temporal coverage and accuracy, within hydrological and hydrodynamic models, with the aim of improving flood forecasting.

The specific objectives of this thesis are:

1 To investigate the effect of StPh sensor locations and different observation accuracies on the assimilation of distributed synchronous streamflow observations in hydrological modelling.
2 To investigate the effect of assimilating distributed uncertain synchronous CS streamflow observations, intermittent in time and space, from StSc sensors in hydrological modelling.
3 To assess the influence of different observational frequencies and related accuracies on the assimilation of asynchronous CS streamflow observations from StSc sensors in hydrological modelling.
4 To investigate the effect of integrating distributed low-cost StSc sensors with a single StPh sensor to assess the improvement in flood prediction performance in hydrological modelling.
5 To study the effect of various DA approaches on the assimilation of synchronous streamflow observations, from existing StPh sensors, in hydraulic modelling.

6 To evaluate the effect of assimilation of synchronous water depth observations, from spatially distributed of StPh sensors, in hydraulic modelling.
7 To assess the integration of distributed StPh, StSc and DySc sensors for assimilation of synchronous CS observations within a cascade of hydrological and hydraulic models.
8 To develop guidelines for using technologies for crowdsourced data assimilation in flood forecasting.

1.5 OUTLINE OF THE THESIS

This thesis is organized into eight chapters. A brief overview of the structure of this thesis is reported below and in Figure 1.9.

Chapter 1 introduces the research background, motivations and objectives of this thesis.

Chapter 2 describes the case studies and their related hydrological and hydraulic models.

Chapter 3 reports the mathematical formulation of various DA methods (direct insertion, nudging scheme, Kalman filtering, ensemble Kalman filtering and asynchronous ensemble Kalman filtering) used in the different models in order to assimilate streamflow and water depth observations from different sensors.

Chapter 4 presents the methods developed for assimilating streamflow observations, synchronous in time, from StPh and StSc sensors, and the related experiments within a semi-distributed model of the Brue catchment. In Experiment 4.1, the effect of StPh sensors' location and their observations' accuracy on the model performances are analysed. In Experiment 4.2 and Experiment 4.3, the influence of variable spatial accuracy and intermittent temporal behaviour of CS observations from StSc sensors on model results are assessed. Finally, in Experiment 4.4, the assimilation of the CS observations coming from an integrated network of StPh and StSc sensors is performed.

Chapter 5 explores the assimilation of asynchronous streamflow observations, from StSc sensors, in hydrological modelling. In particular, the influence of different arrival frequencies and accuracies of the CS observations on the DA performances is explored in Experiment 5.1. In addition, assimilation of CS observations from an integrated network of StPh and StSc sensors is performed to investigate the improvement in the flood prediction performances of an existing EWS in the Bacchiglione catchment (Experiment 5.2).

Chapter 6 investigates the effects of different DA approaches on the assimilation of synchronous streamflow observations from StPh sensors (Experiment 6.1) in a hydraulic model of the Trinity and Sabine Rivers (Texas), and the sensitivity of such models to the assimilation of water depth observations from distributed of StPh sensors in the Bacchiglione River (Experiment 6.2).

Chapter 7 presents a novel application of assimilation of synchronous CS water depth observations, derived from a heterogeneous network of StPh, StSc and DySc sensors into the cascade of hydrological and hydraulic models of the Bacchiglione catchment EWS. In particular, the effects of different citizen engagement levels related to CS observations from StSc and DySc sensors are analysed (Experiment 7.4)in case of.

Chapter 8 describes the conclusions and recommendations for further research.

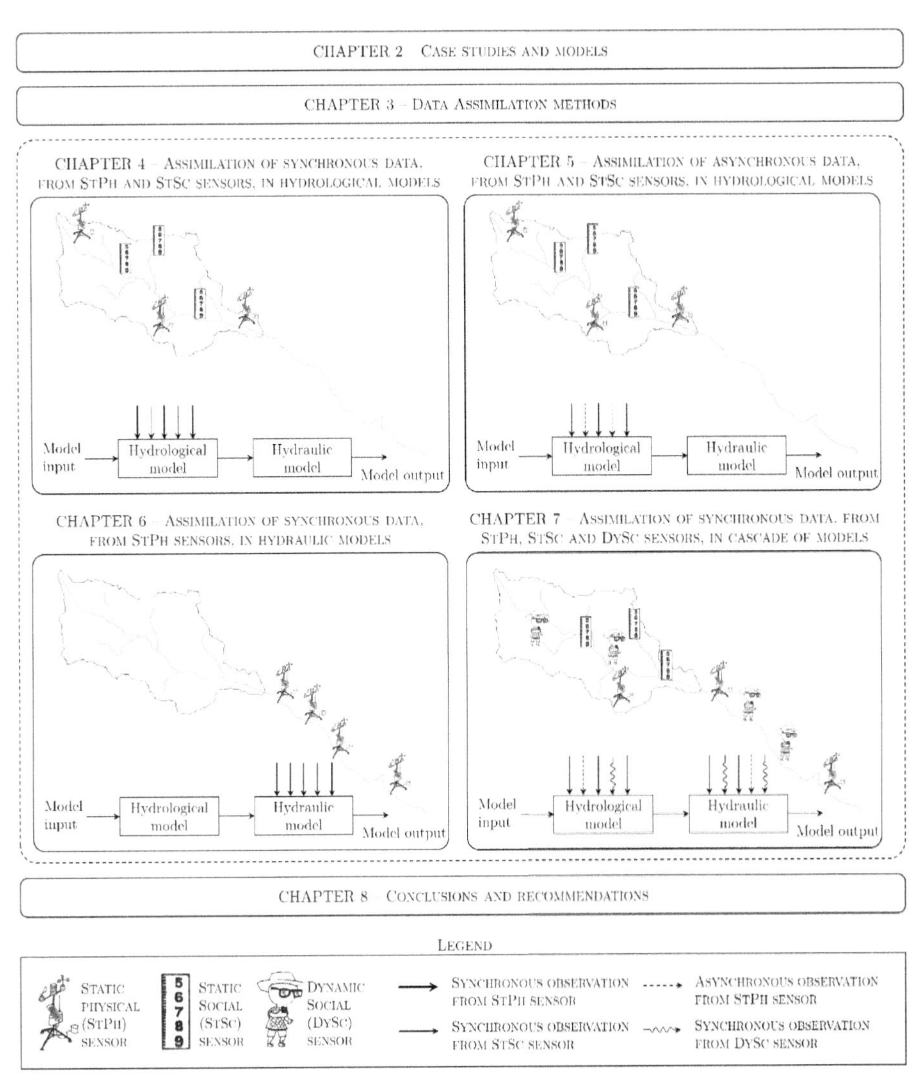

Figure 1.9. Overview of thesis structure

2

CASE STUDIES AND MODELS

The methods developed in Chapters 4 to Chapter 7 of this thesis are applied in three selected case studies with different characteristics. This chapter describes each river basin and the hydrological and/or hydraulic models. The case studies are the Brue catchment (UK), the Bacchiglione catchment (Italy), and the Trinity Sabine rivers (U.S.A.). A lumped and semi-distributed version of a continuous Kalinin-Milyukov-Nash (KMN) cascade hydrological model is applied in order to estimate the flow hydrograph at the outlet section of the catchment. For the Bacchiglione catchment a semi-distributed hydrological and hydraulic model is used to assess water depth values at Vicenza, the outlet of the catchment. A lumped and distributed version of the 3-parameter Muskingum model is implemented along the Trinity and Sabine rivers in order to estimate streamflow values at two particular locations on the rivers and the distributed streamflow values in the Dallas–Fort Worth Metroplex area.

2.1 INTRODUCTION

In this section, description of the main case studies and the implemented hydrological and hydraulic models used in this thesis is provided. The Brue catchment is chosen based because of the availability of rainfall and flow time series within the HYREX project (Moore et al., 2000), while the Bacchiglione basin is one the official cases studies of the WeSenseIt project (Huwald et al., 2013), which is funding this research. In addition, one the main goal of this chapter is also to test the proposed methodology to assimilate crowdsourced observations to then apply it on the existing EWS implemented by Alto Adriatico Water Authority (AAWA) on the Bacchiglione catchment. The Trinity and Sabine River are selected because of the recent flood events occurred in Dallas, TX (USA), and because of the flow data freely available from the National Weather Service (NWS) in USA. Figure 2.1 summaries the hydrological and hydraulic models used in the different case studies, with the indication of the Thesis chapter where the different catchments and model are used.

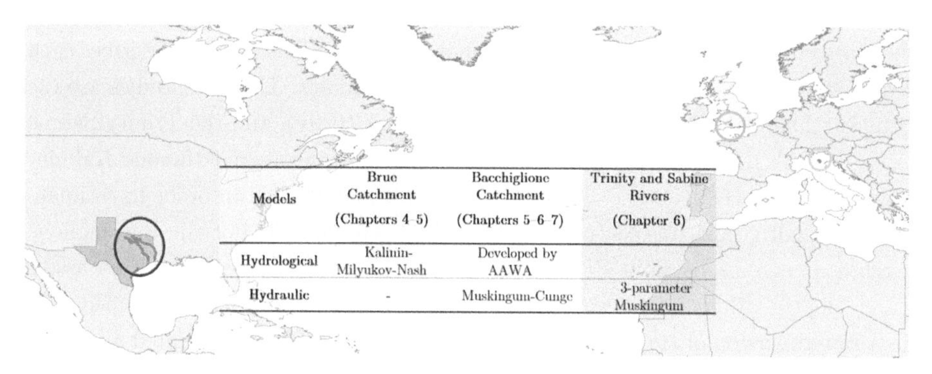

Models	Brue Catchment (Chapters 4-5)	Bacchiglione Catchment (Chapters 5-6-7)	Trinity and Sabine Rivers (Chapter 6)
Hydrological	Kalinin-Milyukov-Nash	Developed by AAWA	-
Hydraulic	-	Muskingum-Cunge	3-parameter Muskingum

Figure 2.1 Summary of the hydrological and hydrodynamic models used in the selected case studies

2.2 CASE 1 - BRUE CATCHMENT (UK)

2.2.1 Catchment description

The Brue catchment is located in Somerset, South West of England, with predominantly rural use and modest slope (Moore et al., 2000). The drainage area of the catchment is about 135 km^2, with time response of about 10 to 12 hours at the catchment outlet, Lovington. Hourly precipitation data are supplied by the British Atmospheric Data Centre from the NERC Hydrological Radar Experiment Dataset (HYREX) project (Moore et al., 2000; Wood et al., 2000) and available at

49 automatic rain stations (Figure 1); average annual rainfall of 867 mm is measured in the period between 1961 and 1990.

Discharge is measured at the catchment outlet by one station at a 15min time step resolution, having an average value of 1.92 m^3/s. For both precipitation and discharge data, a 3-years complete data set, between 1994 and 1996, is available. Discharge observations used come from rating curves that are typically not very accurate (Westerberg et al., 2011). However, in order to evaluate model performances, observed values of discharge at the catchment outlet are assumed to be error-free when compared to model results.

In this thesis, the topography of the area is represented by means of a SRTM 90m resolution DEM which is used to derive the river (streamflow) network and classify it using the approach proposed by Horton (Strahler, 1957).

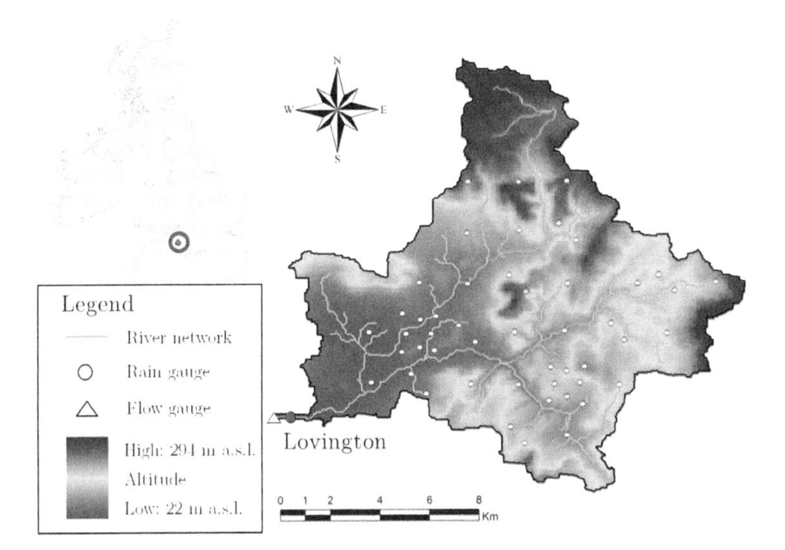

Figure 2.2. Brue catchment and location of the rain and glow gauges

The Brue catchment is selected in this thesis because of the large free available amount of data for both precipitation and streamflow.

2.2.2 Model description

In this section, two different structures of a hydrological model for the Brue catchment are described. (1) A lumped conceptual model, and; (2) A semi-distributed hydrological model. The former is used to assess the effect of assimilating asynchronous CS observations within hydrological models and the latter is used to represent the spatial variability of the uncertain streamflow observations. Below, the description of both models is reported.

Model 1: Hydrological Lumped model

This model is used to develop the methods presented in Chapters 5. The input of the lumped conceptual model is the direct runoff. Such variable is estimated in each sub-catchment using the Soil Conservation Service Curve Number (SCS-CN) method. The CN is calibrated comparing the observed volume of the direct runoff, at the outlet section of the Brue catchment, with the one assessed using the SCS-CN method. The value of *CN* in each sub-catchment is assumed to be equal to the average value of the parameter *CN* within the Brue catchment. In order to estimate the observed quickflow it is necessary to filter the base flow information from the streamflow. To this end, the equation proposed by Eckhardt (2005) is used, in a fashion similar to Corzo and Solomatine (2007), where such equation is employed for building an optimal committee of neural network models to estimate separately base- and excess flow.

For each sub-catchment, a conceptual lumped hydrological model, Continuous Kalinin-Milyukov-Nash (KMN) Cascade, is implemented to estimate the outflow discharge. The KMN model considers a cascade of storage elements (or reservoirs), assuming that the relation between stage, discharge and stored water volume is linear and that the water storage x_t is only a function of the outflow of the reach Q_t (Szilagyi and Szollosi-Nagy, 2010).

$$Q_t = k \cdot x_t \tag{2.1}$$

where k is a time constant that define how fast the water flows out the reservoir. The hydrograph at the catchment outlet can be estimated as convolution of the input I with the impulse-response function h:

$$Q_t = \int_{t_o}^{t} h_\tau \cdot I_{t-\tau} \cdot d\tau = \frac{1}{k(n-1)!} \int_{t_o}^{t} \left(\frac{\tau}{k}\right)^{n-1} e^{-\tau/k} \cdot I_{t-\tau} \cdot d\tau \tag{2.2}$$

where n and k are two model parameters defining the number of storage elements (-) in each sub-catchment and the storage constant (hours) respectively. In this chapter, the parameter k is a linear function between the time of concentration t_c of the given sub-catchment through the equation and a calibration coefficient c:

$$k = c \cdot t_c \tag{2.3}$$

The time of concentration, expressed in hours, is evaluated using the Giandotti equation (Giandotti, 1933), knowing the main topographic characteristics for each sub-catchment. In this chapter the values of the parameters n and c are assumed equal for all sub-catchments. In order to apply the data assimilation techniques, the KMN model is represented as a dynamic state-space system according as

$$\mathbf{x}_t = M(\mathbf{x}_{t-1}, \theta, I_t) + w_t \qquad w_t \sim N(0, \mathbf{S}_t) \tag{2.4}$$

$$\mathbf{z}_t = H(\mathbf{x}_t, \theta) + v_t \qquad v_t \sim N(0, R_t) \tag{2.5}$$

where, $\mathbf{x}_t$ and $\mathbf{x}_{t-1}$ are state vectors at time t and t-1, M is the model operator that propagates the states $\mathbf{x}$ from its previous condition to the new one as a response to the inputs I_t, $\boldsymbol{\theta}$ the set of model parameters, while H is the operator matrix which maps the model states into output $\mathbf{z}_t$. The system and measurements errors w_t and v_t are assumed to be normally distributed with zero mean and covariance $\mathbf{S}$ and $\boldsymbol{R}$, a $n_{obs} \times n_{obs}$ diagonal matrix. In a hydrological modelling system, these states can represent the water stored in the soil (soil moisture, groundwater) or on the earth surface (snow pack). These states are one of the governing factors that determine the hydrograph response to the inputs into the catchment.

In case of the linear systems used in this chapter, the discrete state-space system of Eqs.(2.4) and (2.5) can be represented as follows (Szilagyi and Szollosi-Nagy, 2010):

$$\mathbf{x}_t = \Phi \mathbf{x}_{t-1} + \Gamma I_t + w_t \tag{2.6}$$

$$z_t = \mathbf{H}\mathbf{x}_t + v_t \tag{2.7}$$

where t is the time step, $\mathbf{x}$ is vector of the model states (stored water volume in m^3), $\boldsymbol{\Phi}$ is the state-transition matrix (function of the model parameters n and k), $\boldsymbol{\Gamma}$ is the input-transition matrix, $\mathbf{H}$ is the output matrix, and I and z are the input (forcing) and model output. In the application considered in this thesis, i.e. assimilation of crowdsourcing observations varying in time and space, the matrix R is time dependent since at each time step the error in the measurement is assumed variable, as described in the next sections. The state-transition and input-transition matrixes, and , estimated by Szilagyi and Szollosi-Nagy (2010), are:

$$\Phi = \begin{bmatrix} e^{-\Delta tk} & 0 & 0 & \cdots & 0 \\ \Delta tke^{-\Delta tk} & e^{-\Delta tk} & 0 & \cdots & 0 \\ \frac{(\Delta tk)^2}{2!}e^{-\Delta tk} & \Delta tke^{-\Delta tk} & e^{-\Delta tk} & 0 & \vdots \\ \vdots & \vdots & \ddots & \ddots & 0 \\ \frac{(\Delta tk)^{n-1}}{(n-1)!}e^{-\Delta tk} & \frac{(\Delta tk)^{n-2}}{(n-2)!}e^{-\Delta tk} & \cdots & \Delta tke^{-\Delta tk} & e^{-\Delta tk} \end{bmatrix} \quad (2.8)$$

$$\Gamma = \begin{bmatrix} \left(1-e^{-\Delta tk}\right)\cdot k^{-1} \\ \left[1-e^{-\Delta tk}(1+\Delta tk)\right]\cdot k^{-1} \\ \left[1-e^{-\Delta tk}\left(1+\Delta tk+\frac{(\Delta tk)^2}{2}\right)\right]\cdot k^{-1} \\ \vdots \\ \left(1-e^{-\Delta tk}\sum_{j=0}^{n-1}\frac{(\Delta tk)^j}{j!}\right)\cdot k^{-1} \end{bmatrix} \quad (2.9)$$

where t is the model time step. The matrix **H** is related with the number of storage elements n of the model. For example, for n=3 the matrix **H** is expressed as H=[0 0 k]. The state-transition matrix and the input-transition are reported in (Szilagyi and Szollosi-Nagy, 2010).

Model 2: Semi-distributed hydrological model

A semi-distributed hydrological model is used to assess the flood hydrograph at the outlet section of the Brue catchment and to represent the spatial variability of the uncertain streamflow observations analysed in Chapter 4. For this reason, the Brue catchment is divided into 68 sub-catchments having a small drainage area (on average around 2 km^2). In this way, it is assumed that any observation at a random location in a given sub-catchment would provide the same information content that an observation at the outlet of same sub-catchment (Mazzoleni et al., 2015a).

The lumped model previously described is used in each sub-catchment to calculate the outflow hydrograph. An additional model component is used to propagate the flow of the upstream sub-catchment through the downstream sub-catchment. The propagation of such variable is assessed using the Muskingum channel routing method (Cunge, 1969). The Muskingum model has two parameters, the storage constant K (hours) and the attenuation factor X. The value of K is estimated as a function of the time lag between the two consecutive sub-catchments

$$K = 0.6 \cdot t_c \quad (2.10)$$

where t_c is the concentration time of the downstream sub-catchment z. The attenuation factor X is set to 0.5 for each sub-catchment, which corresponds to the minimum value of flood peak attenuation.

Two different model structures are assumed in this chapter for the semi-distributed hydrological model (see Figure 2.3). The first model structure (MS1) is a sequential one: a sequence of connected sub-catchments where the output from an upstream sub-catchment is used as the input to the following downstream sub-catchment, in addition to the rainfall input. For the second model structure (MS2, common case in most of the semi-distributed models), the output of the upstream sub-catchments is propagated directly to the downstream section and aggregated with the discharge estimated for the downstream sub-catchment; in this case, the sub-catchments are arranged in a parallel structure.

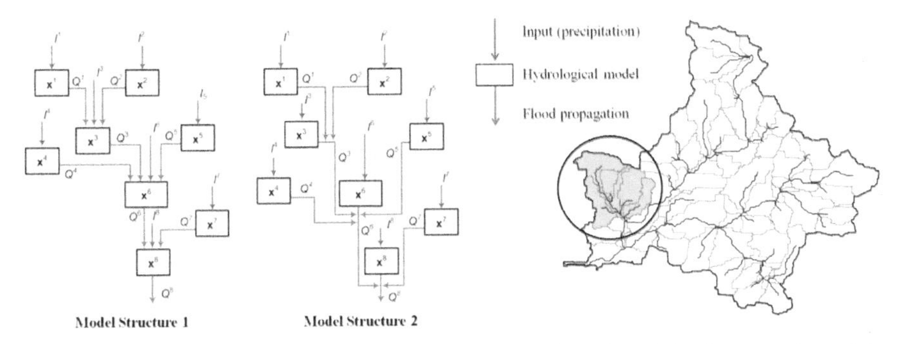

Figure 2.3. Considered structures for the semi-distributed hydrological model

In order to determine average rainfall in each sub-catchment at the necessary resolution, the Ordinary Kriging approach (Journel and Huijbregts, 1978) is employed. This technique allows for optimal interpolation of rainfall data from ground stations (Matheron, 1963) creating a geostatistical model for the whole domain of the precipitation field. The computation of average precipitation is carried out by sampling the precipitation field on fine regular grid (50 x 50 m) within each of the sub-catchments, and then computing a simple average among the sampled points. For the precipitation events used in this paper we assumed isotropy and heteroscedasticity (Savelieva et al., 2008).

Models calibration and validation

In order to calibrate both the lumped and the semi-distributed model, the NSE (Eq.(1.1)) is used as the objective function to compare the observed and simulated value of discharge at the catchment outlet at the time step t. As a result of the calibration procedure, performed using the historical time series of flow, the value of *CN* that minimises the difference between the observed quickflow volume and the direct runoff volume estimated with the SCS-CN method is equal to 87, for both lumped and semi-distributed models. The optimal set of parameters are reported in Table 2.1.

Table 2.1. The optimal set of parameters for the lumped and semi-distributed models

	c	n
Lumped	0.026	4
Semi-distributed (MS1)	1.1	1
Semi-distributed (MS2)	0.8	10

It is worth noting that these parameters values are assumed to be the same for all sub-catchments in case of the semi-distributed model.

Different data sets are used to validate the lumped and semi-distributed hydrological model. In particular, in case of lumped model, five flood events occurred between 28/10/1994 and 16/11/1994 (flood event 1), 08/11/1994 and 29/11/1994 (flood event 2), 24/12/1994 and 10/01/1995 (flood event 3), 16/12/1995 and 30/12/1995 (flood event 4) and 04/12/1995 and 12/12/1995 (flood event 5) are considered. From the results showed in Figure 2.4, it can be seen how well the calibrated lumped model represent the observed flow during flood event 1. For this simple model, which does not take into account complex rainfall-runoff processes, a NSE equal to 0.51 is obtained (see Table 2.2). On the other hand, the lumped model tends to overestimate the observed flow value during the other flood events, leading to negative NSE values. For the semi-distributed model 5 different smaller flood events occurred between 08/11/1994 to 16/11/1994 (flood event A), 28/10/1994 to 07/11/1994 (flood event B), 04/01/1994 to 08/01/1994 (flood event C), 06/12/1994 to 09/12/1994 (flood event D) and from 31/01/1995 to 03/02/1995 (flood event E) are considered. From the validation analysis (Figure 2.5), it can be noticed how for the flood event B the distributed field of precipitation used as input in the model leads to the good results in terms of flow hydrograph at the outlet section of the catchment (see NSE values reported in Table 2.2). On the contrary, during the other events the models tend to underestimate the real observed hydrograph, resulting on NSE values between 0.40 and 0.56 for MS1 and MS2. This can be related to the conceptual nature of the model or to the non-perfect estimation of the average precipitation used as input in each sub-catchment. For both lumped and semi-distributed models, the validation flood events are used for the as hindcasting assimilation of crowdsourced observations.

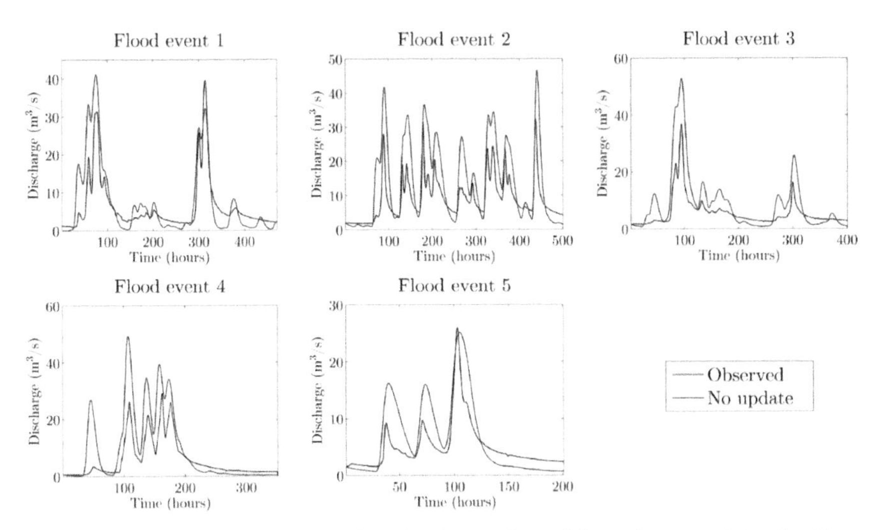

Figure 2.4. Validation results for the lumped model implemented in the Brue catchment

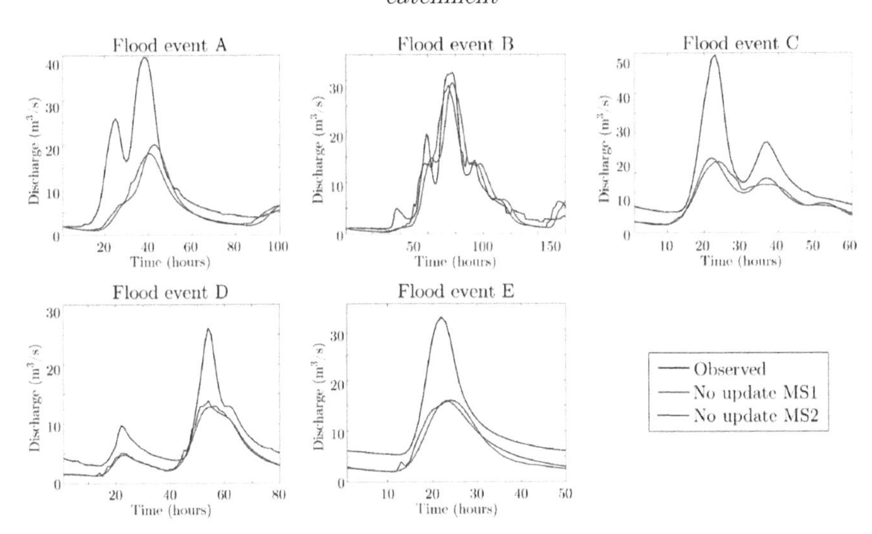

Figure 2.5. Validation results for the semi-distributed model implemented in the Brue catchment, for different model structures MS1 and MS2

Table 2.2. Validation results for the lumped and semi-distributed models

	Lumped	Semi-distr. (MS1)	Semi-distr. (MS2)
Flood event 1	0.51	-	-
Flood event 2	-1.05	-	-
Flood event 3	-0.68	-	-
Flood event 4	-0.53	-	-

Flood event 5	0.13	-	-
Flood event A	-	0.48	0.42
Flood event B	-	0.91	0.87
Flood event C	-	0.40	0.42
Flood event D	-	0.56	0.43
Flood event E	-	0.46	0.51

2.3 CASE 2 - BACCHIGLIONE CATCHMENT (ITALY)

2.3.1 Catchment description

The Bacchiglione River catchment is located in the North-East of Italy, and tributary of the River Brenta which flows into the Adriatic Sea at the South of the Venetian Lagoon and at the North of the River Po delta. The considered area is the upstream part of the Bacchiglione River, which has an overall area of about 400 km^2, river length of about 50 km, river width of 40m and river slope of about 0.5% (Ferri et al., 2012). The main urban area is Vicenza (red point in Figure 2.6), located in the downstream part of the study area. No backwater effects are present. The analysed part of the Bacchiglione River has three main tributaries. On the Western side the confluences with the Bacchiglione are the Leogra and the Orolo River, whose junction is located in the urban area itself. On the Eastern side there is the Timonchio River (see Figure 2.6).

The area supports two important cities, Padua and Vicenza, as well as industrial, agricultural and hydroelectric activities. This is a flood prone area, with recent floods registered during the springs of 2010 and 2013, which affected urbanized areas. Such flood events are generated by more intense and frequent precipitation. For this reason it is important to reduce the flood risk in the area.

The Bacchiglione catchment is selected because it is one of the official case studies of the WeSenseIt Project, which is funding this research. Recently, within the activities of the project (Huwald et al., 2013), 1 StPh sensor and 10 StSc sensors (staff gauges complemented by a QR code, as represented in Figure 1.7), are installed in the Bacchiglione River to measure water level (see Figure 2.7). In particular, the physical sensor is located at the outlet of the Leogra catchment while the three social sensors are located at the Timonchio, Leogra and Orolo catchment outlets respectively (see Figure 2.7). In this context, this thesis will provide an important contribution to the existing EWS implement by Alto Adriatico Water Authority (AAWA), as it will integrate crowdsourced observations and reduce uncertainties in flood prediction.

Hourly information related to rainfall, temperature, wind direction and intensity, humidity, snow, solar radiation, water level are available for the last 12 years and currently used by AAWA for the weather-climate characterization of the case study. A schematization of rain and flow gauges location is reported in Figure 2.6.

This case study is used to develop the methods presented in Chapter 5, 6 and 7.

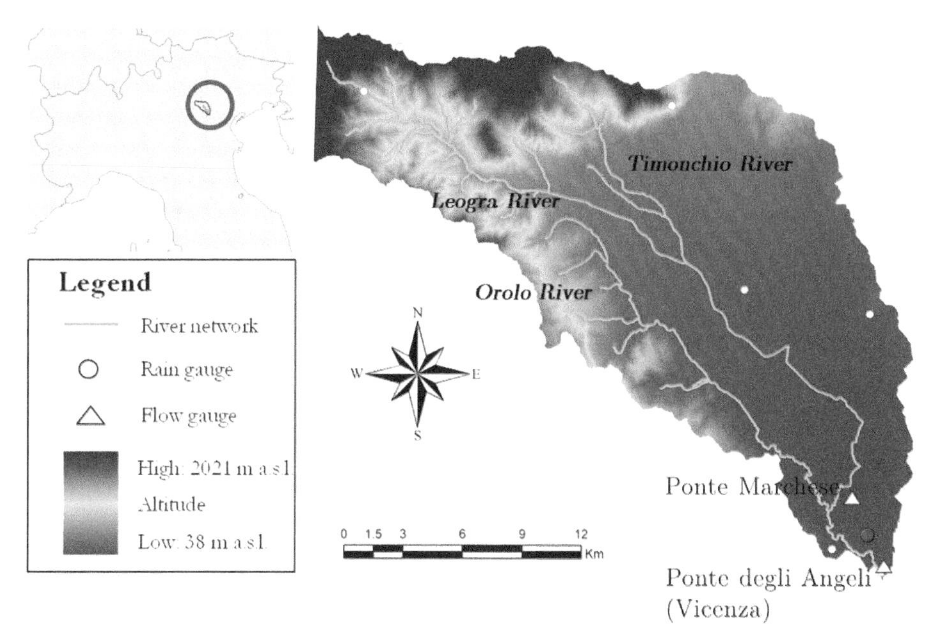

Figure 2.6. Brue catchment with the location of the rain and the historical flow stations of Ponte Marchese (PM) and Ponte degli Angeli (PA)

2.3.2 Model description

After the flood event occurred in November 2010, which hit the territory of Vicenza causing damage to more than one billion euro (Ferri et al., 2012), the Alto Adriatico Water Authority (AAWA) has developed, implemented and made available to the Region of Veneto the operational EWS platform AMICO (Alto adriatico Modello Idrologico e idrauliCO) in order to properly forecast future potential flood events. The platform, already operating on the Bacchiglione catchment, processes the weather and climate data (with a forecast horizon of 3-days), determines the outflows through a geomorpho-climatic model (Rinaldo and Rodriguez-Iturbe, 1996) and estimates the flood wave propagation. The system contains also an automatic optimization module of hydrological model parameters that compares the measured and the simulated discharges at the available measuring points.

In order to represent the distributed hydrological response of this catchment, a semi-distributed hydrological-hydraulic model has been implemented in this thesis (see Figure 2.7).

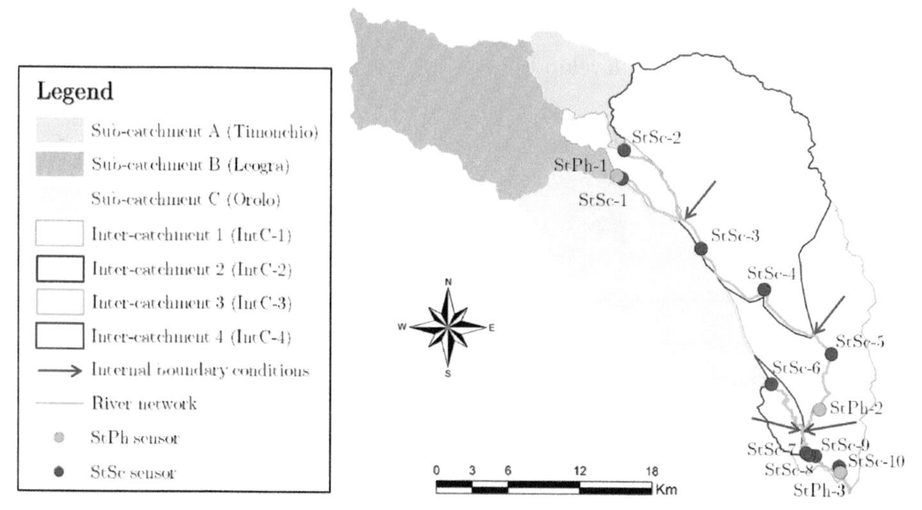

Figure 2.7. Structure of the semi-distributed model, used in this chapter, and location of the StPh and StSc sensors implemented in the catchment by AAWA

In this distributed model, the Bacchiglione catchment is considered as a set of different sub-catchments (seven in this chapter) and the outflow of such sub-catchments is estimated using a hydrological model, developed by the AAWA, which can reproduce the processes of snow accumulation, snowmelt, evapotranspiration, runoff formation and propagation. Moreover, a hydraulic model MIKE11 (DHI, 2005) is connected in cascade to the outlet of the hydrological model in order to estimate the water level in the main river channel and the flood extent in the flood prone area. However, in this thesis, a simplified version of both hydrological and hydraulic models implemented in the AMICO platform are used, which are explained in the next sections. In the schematization of the Bacchiglione catchment (Figure 2.7) the location of StPh and StSc sensors corresponds to the outlet section of three main sub-catchments, Timonchio (A), Leogra (B) and Orolo (C), while the remaining sub-catchments are considered as inter-catchments (used to estimate the lateral inflow to the hydraulic model). For both sub-catchments and inter-catchments, a conceptual hydrological model, described below, is used to estimate the outflow hydrograph. The outflow hydrograph of the three main sub-catchments is considered as upstream boundary conditions of a hydraulic model used to estimate water depth along the river network (blue line in Figure 2.7), while the outflow from the inter-catchment is considered as internal boundary condition.

Hydrological modelling

The hydrological response of the catchment is estimated using the hydrological model developed by AAWA that considers the routines for runoff generation, having precipitation as model forcing, and a simple routing procedure. The processes related to runoff generation (surface, sub-surface and deep flow) are modelled mathematically by applying the water balance to a control volume, of soil depth *D*, representative of the active soil at the sub-catchment scale (see Figure 2.8). The water content *Sw* in the soil is updated at each calculation step *dt* using the following balance equation:

$$Sw_{t+dt} = Sw_t + P_t - R_{sur,t} - R_{sub,t} - L_t - E_t \tag{2.11}$$

where *P* and *ET* are the components of precipitation and evapotranspiration, while R_{sur}, R_{sub} and *L* are the surface runoff, sub-surface runoff and deep percolation model states respectively (see Figure 2.7). In this model, the infiltration *I* is considered as the difference between *P* and R_{sur}. The surface runoff is expressed as:

$$R_{sur,t} = \begin{cases} C \cdot \left(\frac{Sw_t}{Sw_{max}}\right) \cdot P_t, & P_t \leq f = \frac{Sw_{max} \cdot (Sw_{max} - Sw_t)}{Sw_{max} - C \cdot Sw_t} \\ P_t - (Sw_{max} - Sw_t), & P_t \geq f \end{cases} \tag{2.12}$$

where *C* is a calibrated coefficient of soil saturation and Sw_{max} is the water content at saturation point which depends on the CN and the nature of the soil and on its use. The sub-surface flow is considered proportional to the difference between the water content *Sw* at time *t* and the field capacity S_c, equal to the product between *D* and the model parameter θ_c.

$$R_{sub,t} = c \cdot (Sw_t - S_c) \tag{2.13}$$

where *c* is a model parameter. The deep flow is evaluated according to the expression proposed by Laio et al. (2001):

$$L_t = \frac{K_s}{e^{\beta \cdot \left(1 - \frac{S_c}{Sw_{max}}\right)} - 1} \cdot \left(e^{\beta \cdot \left(\frac{Sw_t - S_c}{Sw_{max}}\right)} - 1\right) \tag{2.14}$$

where, K_s is the hydraulic conductivity of the soil in saturation conditions, is a dimensionless exponent characteristic of the size and distribution of pores in the soil. The evaluation of the real evapotranspiration is performed assuming it as a function of the water content in the soil and potential evapotranspiration (PET), calculated using the formulation of Hargreaves and Samani (1982). If Sw_t is bigger than S_{pwp}, i.e. the minimum water content which triggers the evapotranspiration phenomena equal to the product between D and a model parameter θ_w, evapotranspiration is estimated as $ET_t = PET_t \cdot w$, where w is a coefficient of evapotranspiration equal to Sw_t/Sw_{max}. On the other hand, $ET_t = 0$.

Knowing the values of R_{sur}, R_{sub} and L, it is possible to model the surface Q_{sur}, sub-surface Q_{sub} and deep flow Q_g routed contributes according to the conceptual framework of the linear reservoir, see Eqs.(2.6), (2.7), (2.8) and (2.9), at the closing section of the single sub-catchment. In case of Q_{sur}, the stored water volume x_{sur} is estimated using Eq.(2.7) in case of $n=1$, $H=k$ and I_t replaced by R_{sur}.

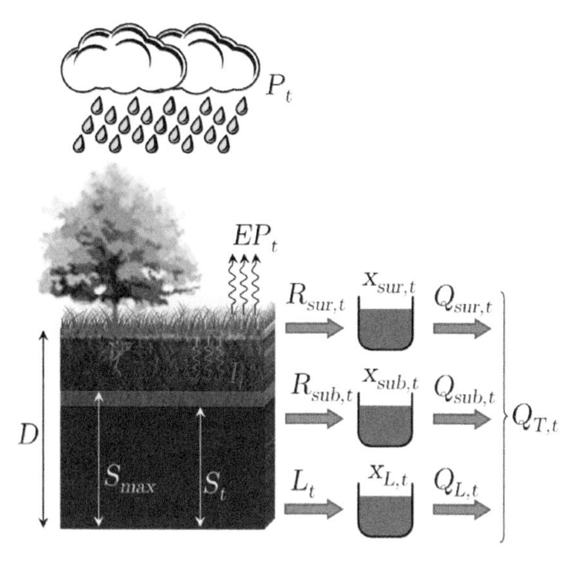

Figure 2.8. Schematization of the main components of the AAWA hydrological model.

In particular, in case of Q_{sur} the value of the parameter k, which is a function of the residence time in the catchment slopes, is estimated relating the slopes velocity of the surface runoff to the average slopes length L. However, one of difficulties involved is proper estimation of the surface velocity, which should be calculated for each flood event (Rinaldo and Rodriguez-Iturbe, 1996). According to Rodríguez-Iturbe et al. (1982), such velocity is a function of the effective rainfall intensity and event duration. In this study, the estimate of the surface velocity is performed suing the relation between velocity and intensity of rainfall excess proposed in Kumar et al. (2002). In this way it is possible to estimate the average time travel and the consequent parameter k. However, such formulation is applied in a lumped way for a given sub-catchment. As reported in McDonnell and Beven (2014) more reliable and distributed models should be used to reproduce the spatial variability of the residence times within the catchment over the time. That is why, in the advanced version of the model implemented by AAWA, in each catchment (e.g. Leogra catchment showed in Figure 2.7) the runoff propagation is carried out according to the geomorphological theory of the hydrologic response considering the overall catchment travel time distributions as nested convolutions of statistically independent travel time distributions along sequentially connected, and objectively identified, smaller sub-catchments. In such model, the parameter k assumes different values for each time step as the rainfall changes. In fact, the variability of residence time is considered according to Rodríguez-Iturbe et al. (1982) by assuming the surface velocity as a function of the effective rainfall intensity (Kumar et al., 2002). In addition, the correct estimation of the residence time should derived considering the latest findings reported in McDonnell and Beven (2014). In case of Q_{sub} and Q_g the value of k is calibrated comparing the observed and simulated discharge at Vicenza as previously described.

In order to find out which model states lead to a maximum increase of the model performance, a preliminary sensitivity analysis is performed. The four model states, Sw, x_{sur}, x_{sub} and x_L, are perturbed by $\pm 20\%$ around the true state value using the uniform distribution, every time step from the initial time step up to the perturbation time (PT).

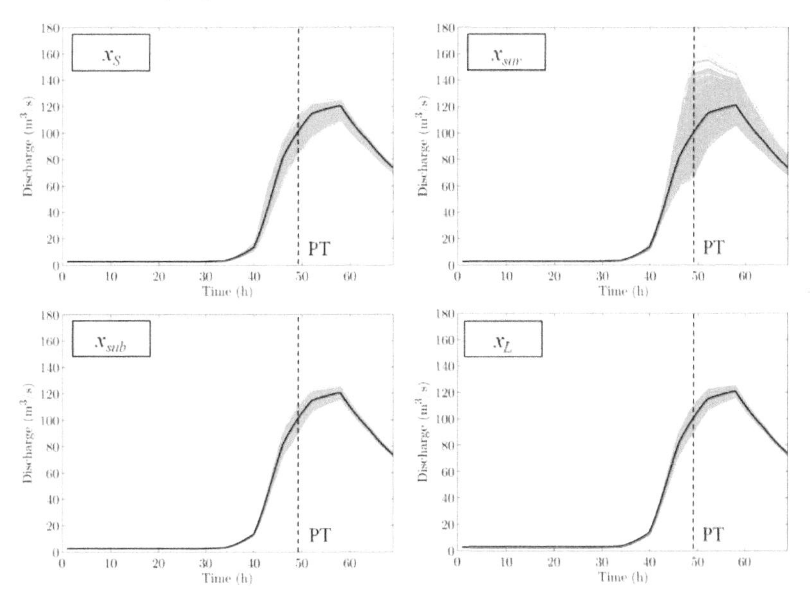

Figure 2.9. Effect of perturbing the model states on the model output, Bacchiglione case study. PT=Perturbation Time.

No correlation between time steps is considered. After PT the model realizations are run without perturbation in order to assess the perturbation effect on the system memory. The results reported in Figure 2.9, in case of the flood event of May 2013, point out that the stored water volume x_{sur} is the most sensitive state if compared to the other five. In addition, the perturbations of all the states seem to affect the model output even after the PT (high system memory). For this reason, only x_{sur} is updated in the DA experiments.

Hydraulic modelling

In the EWS implemented by AAWA in the Bacchiglione catchment, the flood propagation along the main river channel is represented by a one-dimensional hydrodynamic model, MIKE 11 (DHI, 2005). This model solves the Saint Venant Equations in case of unsteady flow based on an implicit finite difference scheme proposed by Abbott and Ionescu (1967).

However, in order to reduce the computational time required by the analysis performed in this chapter to assimilate CS observations, MIKE11 is replaced by a hydrological routing Muskingum-Cunge model (Cunge, 1969; Koussis, 1983; Ponce and Changanti, 1994; Ponce and Lugo, 2001; Todini, 2007). Hydrologic routing models are widely used because of parsimony and minimal data and computational requirements. Due to their simplifying assumptions, however, they are subject to different sources of uncertainty such as input, model parameters and model structures. The derivation of the Muskingum approach is based on the mass balance equation applied to a prismatic river reach between upstream and downstream sections (Todini, 2007).

$$\frac{dS}{dt} = I - O \tag{2.15}$$

where O and I are the is the outflow and inflow discharge, while S is the volume stored in the river reach expressed as linear combination of I and O:

$$S = K \cdot \varepsilon \cdot I + K \cdot (1 - \varepsilon) \cdot O \tag{2.16}$$

where k and ε are two model parameters. Substituting Eq.(2.16) into Eq.(2.15) and solving the system using a centred finite difference approach, the classical derivation of the Muskingum approach can be described as:

$$O_{t+\Delta t} = C_1 I_t + C_2 I_{t+\Delta t} + C_3 O_t \tag{2.17}$$

where Δt is the time step and C_1, C_2 and C_3 are the routing coefficient, constant in time and which sum is equal to 1, which can be calculated from the hydraulic characteristics of the channel reach.

It is worth noting that Eq.(2.17) can be considered as a proper diffusion wave model in case of proper estimation of the model parameter values (Cunge, 1969). Starting from the kinematic routing model, Cunge used a four point time centred scheme to derive the first order kinematic approximation of a diffusion wave model and the consequent new formulation of Eq.(2.17):

$$Q_{t+1}^{j+1} = C_1 Q_t^j + C_2 Q_t^{j+1} + C_3 Q_{t+1}^j \tag{2.18}$$

with

$$\begin{aligned} C_1 &= \frac{c \cdot \Delta t + 2 \cdot \Delta x \cdot \varepsilon}{2 \cdot \Delta x \cdot (1 - \varepsilon) + c \cdot \Delta t} \\ C_2 &= \frac{c \cdot \Delta t - 2 \cdot \Delta x \cdot \varepsilon}{2 \cdot \Delta x \cdot (1 - \varepsilon) + c \cdot \Delta t} \\ C_3 &= \frac{2 \cdot \Delta x \cdot (1 - \varepsilon) - c \cdot \Delta t}{2 \cdot \Delta x \cdot (1 - \varepsilon) + c \cdot \Delta t} \end{aligned} \tag{2.19}$$

where t and j denotes the temporal and spatial discretization, Δx and Δt are spatial and temporal increments, c is the wave celerity and ε is a coefficient equal to:

$$\varepsilon = \frac{1}{2}\left(1 - \frac{q}{c \cdot \Delta x \cdot S_o}\right) \tag{2.20}$$

where q is the unit-width discharge and S_0 is the channel bottom slope. Due to the geometrical characteristics of the river channel in the considered case study, the cross-sections are approximated with rectangular shape and the wave celerity estimated as proposed by Todini (2007):

$$c(h) = \frac{5}{3} \cdot \frac{\sqrt{S_o}}{n} \cdot \left(\frac{A(h)}{P(h)}\right)^{2/3} \cdot \left(1 - \frac{4}{5} \cdot \frac{A(h)}{B(h) \cdot P(h)}\right) \tag{2.21}$$

where n is he manning coefficient while A, B and P are the wetted area, surface width and wetted perimeter respectively.

In order to apply data assimilation, the stochastic state-space form, i.e. in a recursive scheme in which the flow at time step $t+1$ along the river is obtained as a function of flow at time step t of Eq.(2.17) is estimated using the approach proposed by Georgakakos et al. (1990). In such approach, Eq.(2.17) is converted in routing state equation which describes the change in the system vector state **x** responding to the inputs **I** similarly to Eq.(2.6).

$$\mathbf{x}_{t+1} = \Phi \mathbf{x}_t + \Gamma \mathbf{I}_t + w_t \tag{2.22}$$

where $\mathbf{Q}_t=(Q_1^t, Q_2^t,..Q_j^t,..,Q_N^t)$ is the vector of the model states (streamflow in m^3/s) having $n_{state}\times 1$ size, where n_{state} are the number of discrete reaches Δx in which the river is divided, while $\mathbf{I}_t=(Q_L^t, Q_L^{t+1})$ is the 2×1 input vector in which Q_L is the discharge at the upstream boundary condition. In case of lumped model $n_{state}=1$ and Δx is equal to the river length. In Eq.(2.6), w_t denotes the uncertainty of the model structure represented by normal distribution with zero mean and covariance $\mathbf{M_m}$ at the time t, $\mathbf{\Phi}$ ($n_{state}\times n_{state}$) and $\mathbf{\Gamma}$ ($n_{state}\times 2$) represent the state-transition and input-transition matrixes, respectively, derived by Georgakakos et al. (1990) and equal to:

$$\Phi = \begin{bmatrix} C_{1,3} & 0 & \cdots & 0 \\ C_{2,1}+C_{2,2}C_{1,3} & C_{2,3} & \cdots & 0 \\ \vdots & \vdots & \vdots & \vdots \\ \prod_{j=3}^{N=1} C_{j,2}\left(C_{2,1}+C_{2,2}C_{1,3}\right) & \cdots & C_{N-1,3} & \vdots \\ \prod_{j=2}^{N=1} C_{j,2}\left(C_{2,1}+C_{2,2}C_{1,3}\right) & \cdots & \cdots & C_{N,3} \end{bmatrix} \tag{2.23}$$

$$\Gamma = \begin{bmatrix} C_{1,1} & C_{1,2} \\ C_{2,2}C_{1,1} & C_{2,2}C_{1,2} \\ \vdots & \vdots \\ \prod_{j=2}^{N} C_{j,2}C_{1,1} & \prod_{j=2}^{N} C_{j,2}C_{1,2} \end{bmatrix} \tag{2.24}$$

where, for example, $C_{1,3}$ and $C_{1,3}$ are the coefficients C_3 in the first reach Δx and second element of the model states. The associate observation process, which relates the observations to the system states is described as:

$$z_{t+1} = \mathbf{H}_t \mathbf{x}_t + v_t \qquad v_t \sim N(0, \mathbf{R}_t) \tag{2.25}$$

where z_t is a $n_{obs}\times 1$ matrix represents the flow along the river channel at time $t+1$, v is the uncertainty of the measurements represented by normal distribution with zero mean and covariance **R**, while **H** is the $n_{obs}\times n_{states}$ output matrix. Due to the fact that the positions of water depth (WD) observations will change according to the sensor location, also the matrix **H** will be changing accordingly. The Manning's equation will be used to estimate the WD based on observed river cross-sections and compare it with the observed synthetic one in the model performance analyses. In this chapter, the Muskingum-Cunge model is applied to the Bacchiglione catchment, dividing the river network in six main reaches (see Figure 2.10), according to the location of the internal boundary conditions.

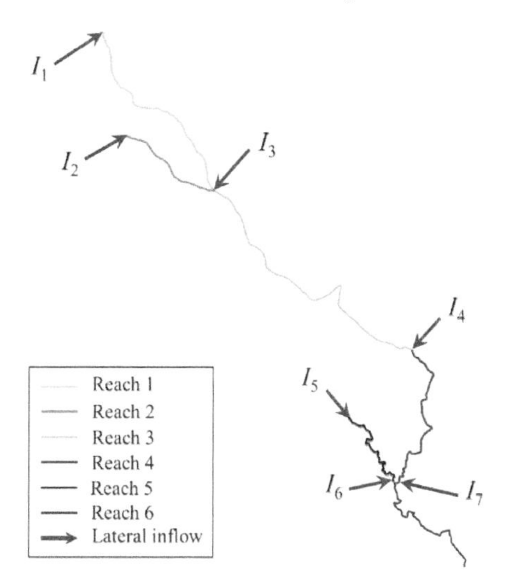

Figure 2.10. Schematization of the different river reaches and relative boundary conditions

For each reach, a distributed formulation of the Muskingum-Cunge model is implemented, with Δx =1000m and Δt =0.9Δx. As it can be noticed, the outputs of the conceptual hydrological model (red arrows), are used as upstream and internal boundary conditions into the proposed Muskingum Cunge model. In Table 2.3 the length of the single reaches of the Bacchiglione catchment network are reported.

Table 2.3. Length, in meters, of each reach in the Bacchiglione catchment network

Reach 1	Reach 2	Reach 3	Reach 4	Reach 5	Reach 6
8000	5000	13000	8000	10000	5000

Model calibration and validation

The calibration of the hydrological model parameters has been performed by AAWA using an adaptation of the "SCE-UA" algorithm (Duan et al., 1992), considering the time series of precipitation from 2000 to 2010, in order to minimise the root mean square error between observed and simulated values of water level at PA (Vicenza) gauged station. In order to stay as close as possible to the EWS implemented by AAWA, the same calibrated model parameters proposed by Ferri et al. (2012) are used in this thesis and reported in Table 2.4.

Table 2.4. Values of the model parameters for each sub-catchments in the Bacchiglione case study

Sub-catchments	Area (m^2)	C (-)	c (-)	θ_c (-)	θ_w (-)	β (-)	Ks (m/h)
Timonchio	20771145.2	0.9	76.9	0.05	0.231	0.158	13.56
Leogra	90266848.1	0.8	76.9	0.05	0.231	0.158	13.56
Orolo	85181159.9	0.5	50	0.001	0.231	0.158	13.56
IntC-1	14025341.5	0.2	55	0.005	0.22	0.133	12.9
IntC-2	118730807.6	0.2	55	0.005	0.22	0.133	12.9
IntC-3	66642409.4	0.32	56	0.001	0.22	0.133	12.9
IntC-4	7923569.02	0.32	56	0.001	0.22	0.133	12.9

In case of the Muskingum-Cunge model, the only parameter that is calibrated in this chapter is the manning coefficient n, used to estimate the WD along the river. In fact, the coefficients C_1, C_2 and C_3 are related to the hydraulic properties of the cross-section, such hydraulic radios and flow celerity. For this reason, an optimal value manning coefficient n, which maximise the NSE value between the observed and simulated rating curve, equal to 0.08 is calculated using the observed rating curve at the gauged station of PA (Vicenza). The semi-distributed hydrological-hydraulic model in the Bacchiglione catchment is validated considering the flood events that occurred in May 2013, November 2014 and February 2016 (see Figure 2.11). Overall, an underestimation of the observed discharge can be observed using forecasted input while the results achieved used measured precipitation tend to well represent the observations. In particular, the flood event of 2013 had high intensity and resulted in several traffic disruptions at various locations upstream Vicenza (Ferri et al., 2012). For flood forecasting, AAWA uses the 3-day weather forecast as the input to the hydrological model.

In Figure 2.12, the observed values water depth, WD, at PA (Vicenza) and PM are compared to the simulated streamflow, Q, and WD values calculated, using MIKE11 and Muskingum-Cunge, considering forecasted (FI) or measured (MI) rainfall data as input in the hydrological model for each sub-catchment during flood event of May 2013 (the most intense among three). The same flood event of May 2013 is used in the next chapter to assimilate crowdsourced water depth observations within the hydrological and hydraulic model.

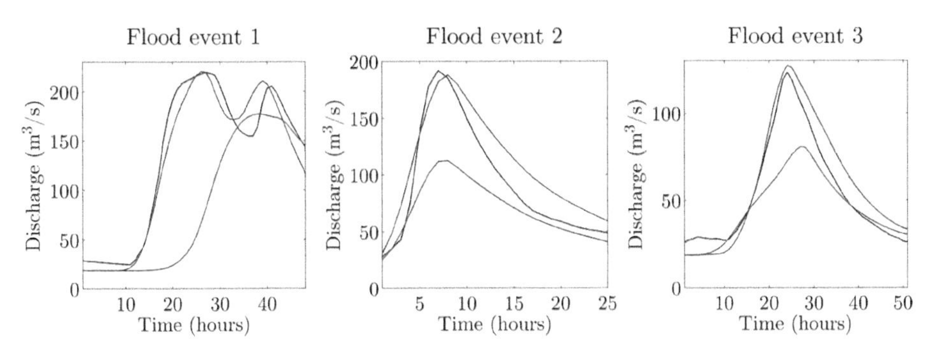

Figure 2.11. The observed (black line) and simulated hydrographs, without update, using measured input (MI, blue line) and forecasted input (FI, red line), for the three considered flood events occurred in 2013 (event 1), 2014 (event 2) and 2016 (event 3) on the Bacchiglione catchment

The hydrographs reported in Figure 2.12 show a good fit between Q and WD, at two different locations, obtained using MIKE11 and the MC model, respectively. However, an overestimation of the WD is observed at PM in both case of measured and forecasted precipitation used in the hydrological model. This can be due to the simplified approach (Manning equation) used to estimate WD and to the way we calibrated the parameter n.

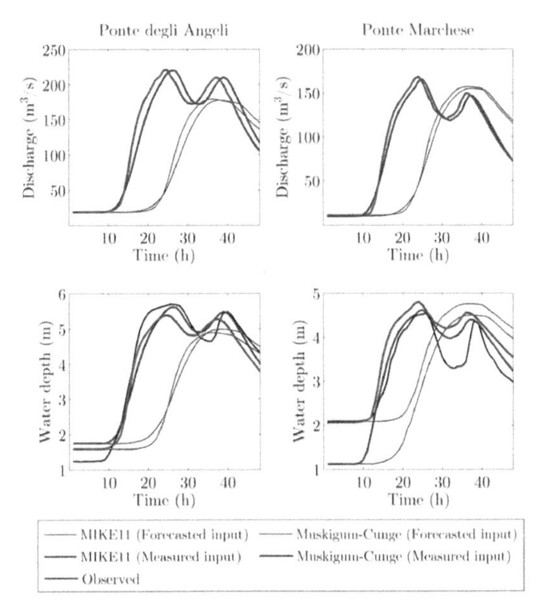

Figure 2.12. Comparison between observed WD value, the results obtained using MIKE11 and the MC model in case of Forecasted precipitation Input (FI) or Measured precipitation Input (MI) during the flood event of May 2013

In fact, the optimal value of n is calculated considering the observed rating curve in Vicenza and not at PM, which might have a slightly different optimal value of n. Observed WD values recorded at the gauging stations of PM and PA are also reported in Figure 2.12.

Figure 2.13 shows that all NSE, R and Bias indicate high correlation and low bias between the results of MIKE11 and MC at PM and PA. High values of NSE and R are obtained for the estimation of WD at PA while low accuracy and high bias are achieved at PM even for values of R above 0.99.

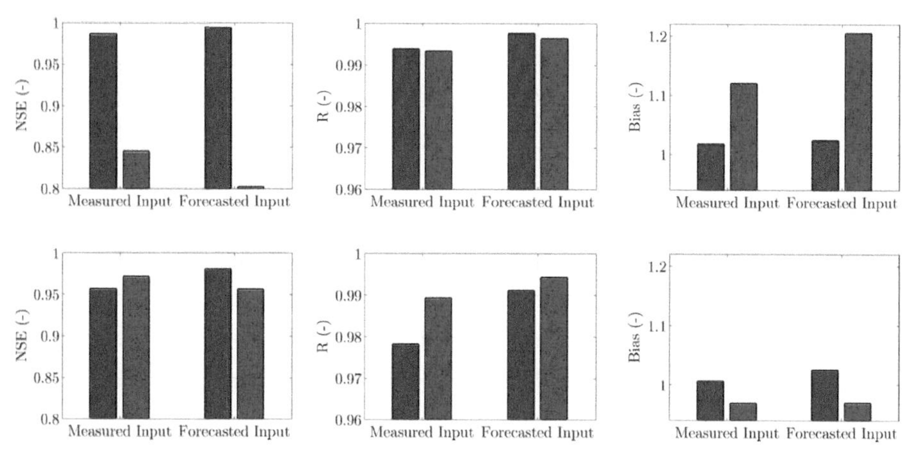

Figure 2.13. Statistical measures obtained comparing results obtained from MIKE11 and Muskingum-Cunge in terms of discharge (blue color) and water depth (red color) at PM (first row) and PA (second row)

2.4 Case 3 - Trinity and Sabine Rivers (USA)

2.4.1 Rivers description

The Trinity River is located in North Texas, U.S.A., it originates from four main forks (Clear, West, Elm and East as showed in Figure 2.14) and it flows south-east up to the Gulf of Mexico. Trinity River is 1140km long, it has a drainage area of about 46500km^2, twenty-one major reservoirs, and its average discharge is about 180m^3/s (USGS, 2016). In its upper part the Trinity River flows though the highly dense urbanized area of the Dallas–Fort Worth (DFW) Metroplex area (4,145,659 inhabitants). Several (32) flow monitoring stations (see Figure 2.14) were installed along the Trinity River and managed by the National Weather Service (NWS).

Data from such stations data are publicly accessible on the internet for a 15-min time step from 2007 to 2015. One the main floods occurred in the 1908 which induced an economic damage of about 5milion dollars and 4000 were left homeless (Barth et al., 2014). Recently, only small flooding occurred, like the one on May and June 2015 (see Figure 2.15).

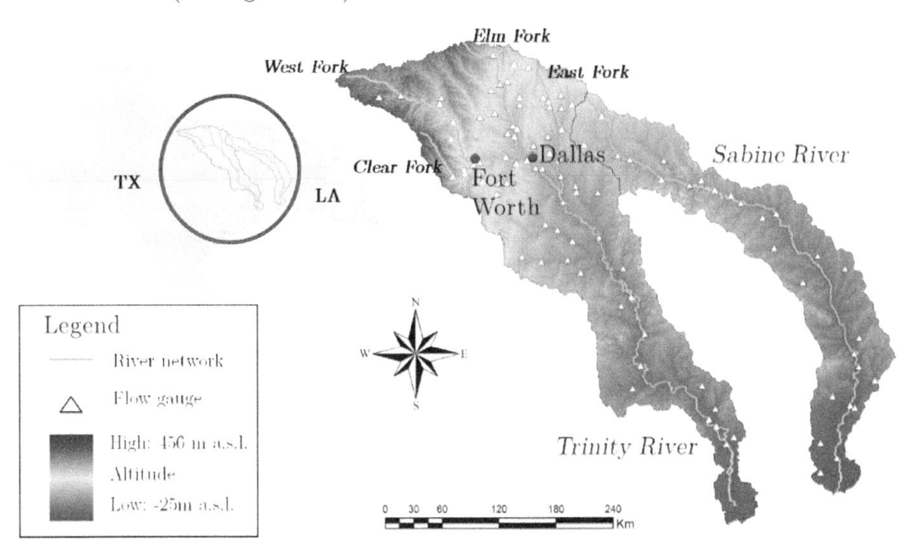

Figure 2.14. Trinity and Sabine Rivers with the location of the NWS StPh flow stations.

Figure 2.15. Images of the 2015 flood event that affected Dallas (left side, source Dallas Morning News) and near Eastern end of Grand Praire/Western end of Dallas (Source WFAA Dallas).

The Sabine River is a transboundary River between Texas, TX, and Louisiana, LA (see Figure 2.14). It has a drainage area of about 25270 km^2 of which 19230 km^2 (76%) in Texas and 6040 km^2 (24%) in Louisiana (Phillips, 2008). Sabine River flows from the Blackland Prairie east of Dallas to the western Gulf Coastal Plain (Heitmuller, 2014). In the recent flood event of January 2016, the Sabine River (in the East Texas side) was hit by a month-long flood. Different citizens could not access their houses for several days due to the duration of the flood event.

This case study is used to develop the methods presented in Chapter 6.

2.4.2 Model description

In case of the Trinity and Sabine Rivers, a 3-parameter Muskingum model (O'Donnell, 1985) is implemented in order to represent the flood propagation. In the implementation of the basic Muskingum model, reported in the previous section, no lateral inflows or outflows along the reach are considered. Instead, if a substantial tributary is present, the routing process can be terminated at the confluence, augmenting the main channel flow by the tributary for the next reach (O'Donnell, 1985). However, if the lateral flow is uniformly distributed, the approach proposed by Georgakakos et al. (1990), which includes an additional term to Eq.(2.17) can be used.

$$\mathbf{x}_{t+1} = \Phi \mathbf{x}_t + \Gamma \mathbf{I}_t + \Lambda \mathbf{q}_t + w_t \tag{2.26}$$

where $\mathbf{x}$ is the model states matrix (streamflow), $\mathbf{q}$ is the later unit inflow or outflow along the reach Δx during the interval Δt. The matrix $\mathbf{\Lambda}$ and the coefficient C_4 can be estimated as

$$\Lambda = \begin{bmatrix} C_{1,4} \\ C_{2,4} + C_{2,2}C_{1,4} \\ \vdots \\ C_{N,4} + C_{N,2}C_{N-1,4} + \cdots + \prod_{j=2}^{N} C_{j,2}C_{1,4} \end{bmatrix} \tag{2.27}$$

$$C_4 = \frac{2 \cdot q \cdot c \cdot \Delta t \cdot \Delta x}{c \cdot \Delta t + 2 \cdot \Delta x \cdot (1 - \varepsilon)} \tag{2.28}$$

O'Donnell (1985) presented a direct and efficient method, which will be used in this chapter, to extend the basic 2-parameter Muskingum model to a 3-parameter model employing a simple assumption about later inflow along the river reach. In fact, O'Donnell (1985) assumed that the lateral inflow is directly proportional, through a coefficient α, to the inflow into the reach. In this way, the term I in Eqs.(2.15) and (2.16) can be written as $I(1+\alpha)$ and, consequently, the coefficients d_1, d_2 and d_3 can be estimated as function of the coefficient C_1, C_2 and C_3 previously estimated:

$$\begin{aligned} d_1 &= (1+\alpha)\cdot C_1 \\ d_2 &= (1-\alpha)\cdot C_2 \\ d_3 &= C_3 \end{aligned} \tag{2.29}$$

Obviously, if there is no lateral inflow the coefficient α are equal to zero and d_1, d_2 and d_3 will coincide with the coefficient C_1, C_2 and C_3 calculated in Eq.(2.19). The procedure used to estimate the state matrix and consequent flow along the reach coincides with the one reported in the previous section. However, in this case, the coefficient d_1, d_2 and d_3 are considered constant in time and function only of the parameters X, K and α.

For the purposes of this chapter, only two reaches of the Trinity River (A and C in Figure 2.16) and one for the Sabine River (B in Figure 2.16) are considered: (A) middle part of the Trinity River between the RSRT2 upstream station and the TDDT2 station used to validate model results; (B) lower part of the Sabine River between the BWRT2 upstream station and the DWYT2 downstream station; (C) upper Trinity River part, enclosed in the DFW area, which is divided into the sub-reach C_1 (FWOT2 as upstream boundary condition), sub-reach C_2 (CART2 as upstream boundary condition) and sub-reach C_3 (DALT2 station at Dallas used to compare observed and simulated streamflow). The confluences of sub-reach C_1 and C_2 are used as upstream boundary conditions for sub-reach C_3 (see Table 2.5).

Table 2.5. Upstream and downstream stations of the different reaches along the Trinity and Sabine Rivers considered in this Thesis

	Upstream station	**Downstream station**
Reach A	RSRT2	TDDT2
Reach B	BWRT2	DWYT2
Reach C	FWOT2 (sub-reach C_1)	DALT2 (sub-reach C_3)
	CART2 (sub-reach C_2)	

In case of reaches A and B, a lumped version of the 3-parameter Muskingum with $\Delta x = L_{reach}$ and $n_{state}=1$ is used. On the contrary, a distributed hydraulic model with Δx=1000m and $\Delta t = 0.9 \cdot \Delta x$ (as in case of the Bacchiglione catchment) is implemented in reach C. The main advantage of a distributed formulation over a lumped one is that is it possible to estimate flow characteristics at different points along the reach of interest. For this chapter the estimation of flow values at particular target points within the DFW area is important, as well as to assimilate crowdsourced observations coming from urbanized areas in order to better predict future flood situations and reduce the consequent economic damages.

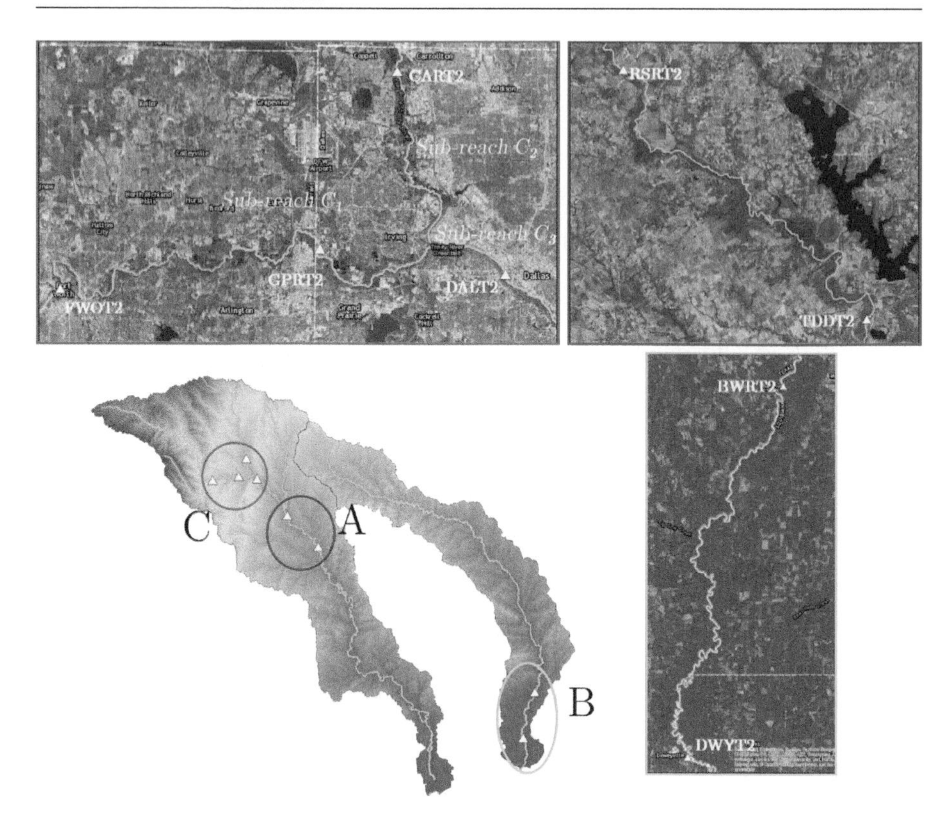

Figure 2.16. Trinity and Sabine Rivers with the location of the NWS StPh flow stations.

Table 2.6. Length, in meters, of each reach in the Trinity and Sabine Rivers

Reach A	Reach B	Sub-reach C_1	Sub-reach C_2	Sub-reach C_3
78000	88000	76000	30000	6000

Model calibration and validation

Calibration for the hydraulic model implemented in the reaches A and B is performed in Lee et al. (2011b), by means of the least squares minimization technique using the Broyden-Fletcher-Goldfarb-Shanno variant of Davidon-Fletcher-Powell minimization (DFPMIN) algorithm (Press et al., 1992). The DFPMIN algorithm implements the quasi-Newton method. In case of reach C, a genetic algorithm (GA) approach (Deb et al., 2002) is used to calibrate the parameters K, X and α values, assuming the same for the three sub-reaches and using the flow time series between 01/10/2007 and 01/10/2013. The optimal parameters values estimated for the three reaches in the Trinity and Sabine Rivers are reported in Table 2.7

Table 2.7. Optimal parameters values for the 3-parameter Muskingum model implemented along reaches A, B and C

Parameters	Reach A	Reach B	Reach C
K	47.28	78.97	612.01
X	0.47	0.35	0.12
a	0.10915	0.10460	0.00588

The validation of the lumped hydraulic model is performed comparing a 5-years long time series (from 01/01/2002 to 31/12/2006) as well as a 12-years long time series (from 01/01/1996 till 31/01/2007) of observed and simulated flows at the TDDT2 and DWYT2 stations for the reach A and B respectively (see Figure 2.17 and Figure 2.18). Six different flood events, having high intensity and long duration, are considered for the river reaches.

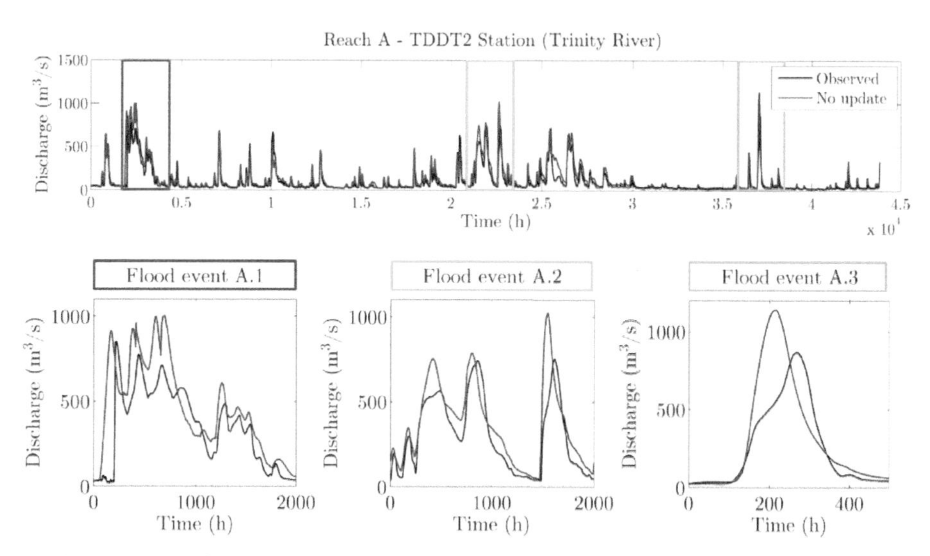

Figure 2.17. Comparison between observed and simulated flow values at the TDDT2 station, in reach A, for the total validation period and 3 single flood events.

In case of reach A, three flood events occurred from 16/03/2002 to 08/06/2002 (flood event A.1), from 29/05/2004 to 20/08/2004 (flood event A.2), and from 14/03/2006 to 04/04/2006 (flood event A.3), are considered and shown in Figure 2.17.

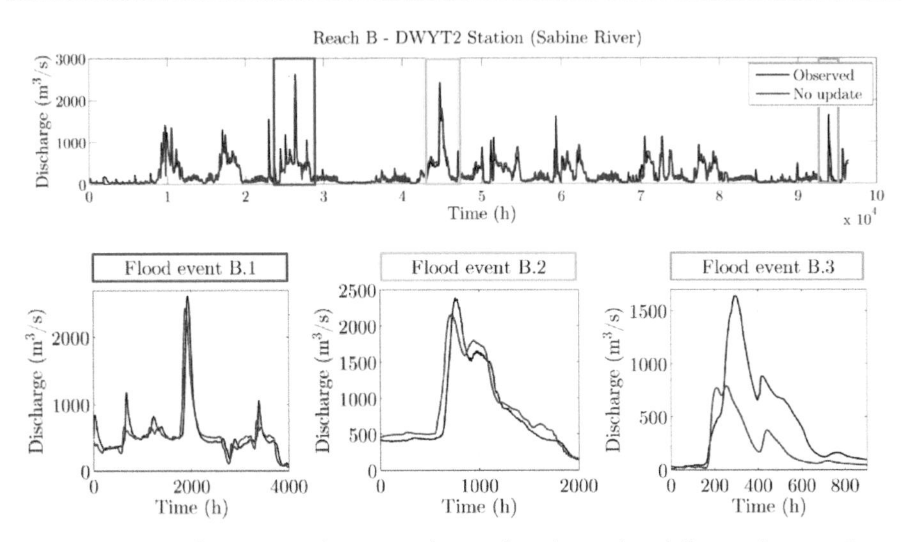

Figure 2.18. Comparison between observed and simulated flow values at the DWYT2 station, in reach B, for the total validation period and 3 single flood events.

On the other hand, in case of reach B the analysis are focused on the flood events that took place between 17/10/1998 and 02/04/1999 (flood event B.1), from 07/01/2001 to 31/03/2001 (flood event B.2), and from 09/09/2006 to 16/10/2006 (flood event B.3), as reported in Figure 2.18. See Table 2.8 for a summary of the considered events.

Table 2.8. Summary of the events used in the simulations on reaches A and B

	Flood event *.1	**Flood event *.2**	**Flood event *.3**
Reach A	16/03/2002 - 08/06/2002	29/05/2004 - 20/08/2004	14/03/2006 - 04/04/2006
Reach B	17/10/1998 - 02/04/1999	07/01/2001 - 31/03/2001	09/09/2006 - 16/10/2006

The results in Figure 2.17 show a systematic overestimation of the simulated flow from the model in reach A, while in case of reach B the model tend to anticipate and underestimate the flow peaks for the considered flood events. . Regarding the distributed 3-parameter Muskingum, the recent flood event occurred between 12/05/2015 and 01/08/2015 which affected the urbanized area of DFW is considered. The results reported in Figure 2.19 show an overestimation (NSE=0.63) of the observed flow at the DALT2 station. Peak 1, reported in Figure 2.19, is used to compare predicted and observed streamflow hydrographs due to the overestimated representation of simulated flow in such situation.

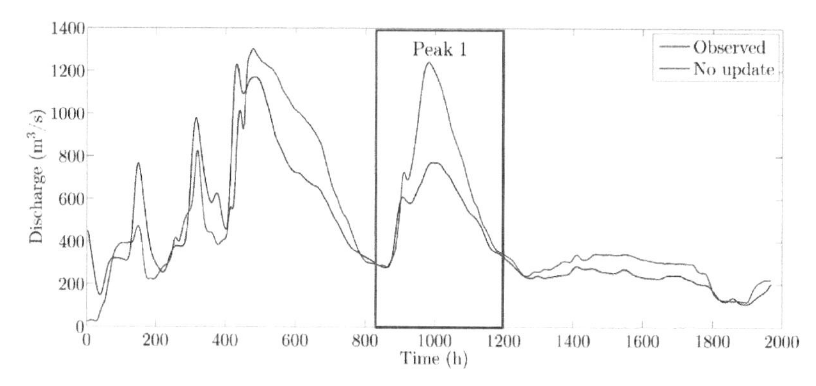

Figure 2.19. Comparison between observed and simulated flow values at the TDDT2 and DWYT2 stations along reach A and B for the total validation period and 2 flood events

2.5 CASE 4 - SYNTHETIC RIVER REACH

A synthetic river with pre-defined cross-sections is used in Chapter 6 to test the assimilation of water level observations and the effect of model and boundary error definition into a Muskingum-Cunge model. For this purpose, a rectangular channel with an increasing width B(m) equal to $50+S_f{}^*x$, where x is the distance along the river and S_f the bed slope, a Manning's coefficient n of 0.035, and a total channel length L equal to 50 km is considered. In this river reach dx is set equal to 1000m while dt is 0.9^*dx in order to achieve Courant values smaller than 1 and consequent stability of the solution. Upstream boundary condition of the synthetic river are assumed to coincide with two flood events occurred in Brue River between 08/11/1994 and 16/11/1994 (flood event A) and from 28/10/1994 to 07/11/1994 (Flood Event B).

3

DATA ASSIMILATION METHODS

This chapter illustrates the data assimilation (DA) methods used in this thesis to integrate uncertain distributed crowdsourced observations from different types of sensors within the hydrological and hydraulic models implemented in the various case studies described in Chapter 2. In this thesis, the standard versions of the DA methods are implemented. However, in order to account for the random nature of crowdsourced observations, a particular definition of the observational error is provided and discussed in detail in the next chapters.

3.1 INTRODUCTION

Model updating methods are becoming important tools in hydrology for integrating the real-time hydrological observations into water system models and thus reducing uncertainty in flood prediction (Liu et al., 2012). The hydrological and hydrodynamic models utilize input variables, that are either measured or estimated (e.g. areal precipitation, air temperature, potential evapotranspiration), into a set of equations that contain state variables and parameters (Refsgaard, 1997). Typically, the parameters main constant while the state variables vary in time. Model output, in most of the cases discharge or water level, are observables and they can be used in real time. Model updating methods allow for either update model input, states, parameters or outputs (Figure 3.1) as new observations become available (Refsgaard, 1997; WMO, 1992). It is worth noting that, as Refsgaard (1997) stated, usually the processes previously described are denoted as Data Assimilation (DA). However, in this Thesis, DA methods are referred to a particular group of model updating methods in which only model states are updated (Lahoz et al., 2010).

a) **Update of input variables:** In water system modelling, inputs are usually considered as dominating sources of uncertainty. Update input variables such as precipitation or temperature can be used in operational forecasting models to reduce model uncertainty (Alberoni et al., 2005). This update method can be manually performed by modellers (e.g. Bergström, 1991) or automatic (Sittner and Krouse, 1979).

b) **Update of state variables:** A wrong state initialization might lead to an erroneous estimation of the model states (e.g. soil moisture) and consequent output. For this reason, state estimation methods as DA are important tools, applied also in hydrology, in order to efficiently use observational information, reliable or scarce, to find the best new estimate of the dynamic models states and improve model predictions (Robinson et al., 1998; McLaughlin, 1995; Refsgaard, 1997; McLaughlin, 2002; Madsen and Skotner, 2005) and to reduce modelling uncertainty (WMO, 1992). Walker and Houser (2005) gave a brief overview about the history of hydrological data assimilation and discussed different assimilation methods. Recently, Liu et al. (2012) presented a comprehensive literature review about the latest advanced of data assimilation procedures in operational flood forecasting by means of water observations coming from physical sensors or remote sensing in a distributed fashion.

c) **Update of model parameters**: Correction of model parameters, originally assessed through optimal calibration using historical data can be difficult to perform due to large number of parameters involved. In particular, Kachroo (1992) stated that: "*It is intuitively difficult to accept that the operation of any hydrological system can change significantly over such a short interval of time as the observation interval. Therefore, recalibrating the model at every time step has no real advantages, other than perhaps some computational attraction and that only when applied to simple forecasting models of the system analysis type*". For these reasons, updating of model parameters is less common than the other three types of updating in flood forecasting (Young, 1984; Xie and Zhang, 2010; Lü et al., 2013). Examples of combined updating of model states and parameters are reported in Moradkhani et al. (2005a).

d) **Update of output variables:** The output updating method (or error prediction) is based on the fact that the errors between the model predictions and the new observations are usually found to be highly correlated. For this reason, regressive model such as an auto-regressive moving average (ARMA) models (Box et al., 1970) can be used to forecast the future error values and improve model predictions. Earlier studies of error prediction have been reported by Jamieson et al. (1973), Lundberg (1982), Babovic et al. (2001) and Abebe and Price (2004).

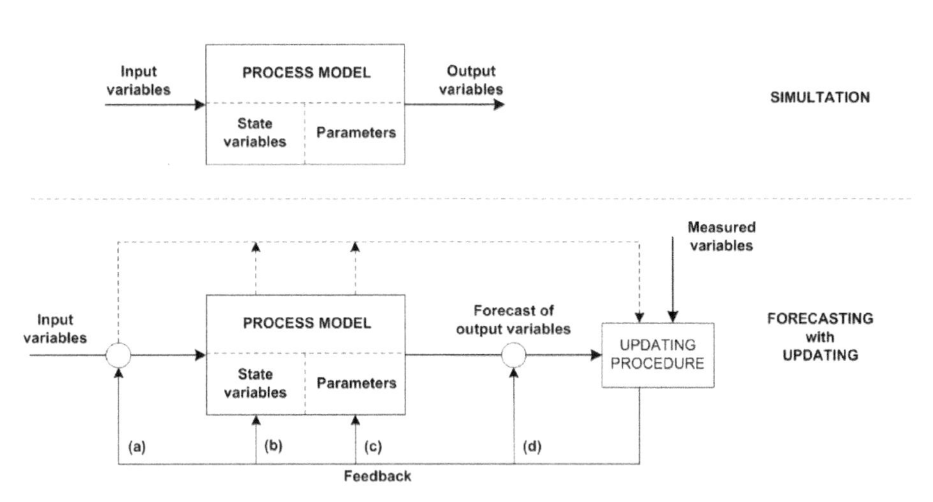

Figure 3.1. Schematic diagrams of simulation and forecasting with emphasis on the four different updating methodologies (modified after WMO 1992)

The focus of this thesis is mainly on the use of DA techniques for integrating CS observations within hydrological and hydraulic model. This chapter does not aim to propose a new DA approach but instead, use the already existing ones and adapt them to the different needs in order to account for the random spatial and temporal characteristics of CS observations.

First of all it is necessary to introduce some terminology used in data assimilation (Walker and Houser, 2005). In this thesis, only sequential DA methods are used. In particular, direct insertion (DI), Nudging scheme (NS), Kalman Filter (KF), Ensemble Kalman filter (EnKF) and Asynchronous Ensemble Kalman filter (AEnKF) are used and described below.

3.2 Direct insertion

In Direct Insertion (DI), the model states are directly replaced with the observations, whenever available.

$$\mathbf{x}_{t,l}^{+} = z_t^o \quad (3.1)$$

Where z^o and x are the observation value and the updated model state at the time step t at a particular location l. The superscript + indicates the state value. The statistical hypothesis of this method is that data measurements are exact and reliable while the mathematical model is wrong (Refsgaard, 1997). However, the risk of this approach is that unbalanced state estimates may result, which causes model shocks (Walker et al., 2001).

3.3 Nudging scheme

Nudging scheme (NS) technique consists on an adding a term to sum in the update model equation in order to "force" the model state closer to the observations. In this approach the nudging term is proportional to the difference between the model states, at a given grid point and time step, and the observations calculated at the corresponding grid point, i.e. data residual. Such term is varying in time but cannot be too large to avoid model disruptions. The general formulation of the NS is:

$$\mathbf{x}_{t+1} = F(\mathbf{x}, \theta, t) + Kn_t \cdot (z_t^o - \mathbf{x}_t) \quad (3.2)$$

where K_n is the nudging (or gain) matrix and F is the model forcing terms (Auroux et al., 2008). Stauffer and Seaman (1990) and Houser et al. (1998) proposed an approach to nudge the model towards regularly spaced observations, or towards randomly spaced observations during a period of time and space.

$$\mathbf{x}_{t+1} = F(\mathbf{x}, \theta, t) + Kn_t \cdot \frac{\left[\sum_{z=1}^{N_{obs}} Wn_z^2(\theta, t) \cdot \gamma_z \cdot (z_t^o - \mathbf{x}_t)_z\right]}{\sum_{z=1}^{N_{obs}} Wn_z^2(\theta, t)} \tag{3.3}$$

Where Wn is the weighting function that specify the temporal and spatial variability of the observation z, N_{obs} is the total number of observations assimilated, γ is the observation quality factor that can vary between 0 and 1 and accounts for characteristic errors in measurement systems and representativeness, while x^t_o is the observation at the model grid. However, the approach used by Brocca et al. (2010) is used in this chapter to assimilate streamflow observations, at a given location l, to update hydraulic model state variables $\mathbf{x}$, i.e. flow, as:

$$\mathbf{x}_t^+ = \mathbf{x}_t^- + Kn_t \cdot (z_t^o - \mathbf{x}_t^-) \tag{3.4}$$

where Kn is estimated as:

$$Kn_t^+ = \frac{S_t}{S_t + R_t} \tag{3.5}$$

where S_t and R_t are the model and observational error variance at the time step t, defined in Eqs.(2.4) and (2.5). The superscript – indicates the forecasted matrices, while the superscript + indicates the updated matrices. In case of perfect measurement R_t is equal to 0, and consequently Kn equal to 1, obtaining $\boldsymbol{x}^+=z^o$ like in the previous case of direct insertion. On the other hand, if the model is assumed perfect S_t is equal to 0, Kn is equal to 0, and this means that there will be no update since $\mathbf{x}^+=\mathbf{x}^-$. Although this DA approach it is not statistically optimal (Brocca et al., 2010), it can be used to assimilate CS observations of flow due to is low computational time costs. A detailed review of nudging methods is proposed by Park and Xu (2013).

3.4 Kalman Filter

Kalman filtering theory is one of the most used approaches when new hydrological observations are available (Robinson et al., 1998; Heemink and Segers, 2002; McLaughlin, 2002; Moradkhani et al., 2005b; Walker and Houser, 2005; Liu and Gupta, 2007; Reichle et al., 2008). In particular, Kalman Filter (KF, Kalman, 1960) is a mathematical tool which allows estimating, in an efficient optimal recursive way, the state of a process governed by a linear stochastic difference equation as response of real-time (noisy) observations. KF is optimal under the assumption that the error in the process is Gaussian; in this case KF is derived by minimizing the variance of the system error (state error) assuming that the model state estimate is unbiased. KF is recursive because it updates model states considering only the last available observation, without requiring all previous data to be kept in storage and reprocessed every time step, allowing for a faster computation. However, KF can be applied only in case of linear dynamic systems. In an attempt to overcome these limitations, various variants of the Kalman filter, such as the extended Kalman filter (EKF), unscented Kalman filter and ensemble Kalman filter (EnKF) have been proposed. Kalman filter procedure can be divided in two steps, namely forecast (background) equations, Eqs.(3.6) and (3.7),

$$\mathbf{x}_t^- = \Phi \mathbf{x}_{t-1}^+ + \Gamma I_t + w_t \tag{3.6}$$

$$\mathbf{P}_t^- = \Phi \mathbf{P}_{t-1}^+ \Phi^T + \mathbf{S}_t \tag{3.7}$$

and update (or analysis) equations Eqs.(3.8), (3.9) and (3.10):

$$\mathbf{K}_t = \frac{\mathbf{P}_t^- \mathbf{H}^\mathrm{T}}{\mathbf{H}\mathbf{P}_t^- \mathbf{H}^\mathrm{T} + R_t} \tag{3.8}$$

$$\mathbf{x}_t^+ = \mathbf{x}_t^- + \mathbf{K}_t \cdot (z_t^o - \mathbf{H}\mathbf{x}_t^-) \tag{3.9}$$

$$\mathbf{P}_t^+ = (\mathbf{I} - \mathbf{K}_t \mathbf{H})\mathbf{P}_t^- \tag{3.10}$$

where $\mathbf{x}$ is the $n_{state} \times 1$ state matrix, $\mathbf{K}_t$ is the $n_{states} \times n_{obs}$ Kalman gain matrix, $\mathbf{P}$ is the $n_{states} \times n_{states}$ error covariance matrix, z^0 is the new observation and $\mathbf{M}_Q$ is the model error matrix. $\mathbf{\Phi}$ and $\mathbf{\Gamma}$ represent the state-transition and input-transition matrices, which change according to the model type and structure. In case of application of KF to a Muskingum-Cunge model, the covariance matrix $\mathbf{P}$ in Eq.(*3.7*) accounts the errors in the boundary conditions and is estimated as:

$$\mathbf{P}_t^- = \Phi \mathbf{P}_{t-1}^+ \Phi^T + \mathbf{\Gamma} \mathbf{M}_\mathrm{b} \mathbf{\Gamma}^\mathrm{T} + \mathbf{S}_t \tag{3.11}$$

where $\mathbf{M}_\mathrm{b}$ is the 2×2 covariance matrix of the boundary conditions matrix $\mathbf{I}$ in Eq.(2.22). In this chapter, the calculation of $\mathbf{\Phi}$ and $\mathbf{\Gamma}$ is described in Chapter 3 for the different model used. The prior model states $\mathbf{x}$ at time t are updated, as the response to the new available observations, using the analysis equations Eqs.(3.8) to (3.10). This allows for estimation of the updated states values (with superscript

+) and then assessing the background estimates (with superscript –) for the next time step using the time update equations Eqs.(3.6) and (3.7).

A key issue in the implementation of the Kalman filter is the determination of model errors. In fact, an overestimation of model errors can reduce the confidence in the model, and thus the KF would overly rely on observations (Sun et al., 2015). On the other hand, an underestimation of model errors might increase the trust in the model, discarding the information from the new observations (Kitanidis and Bras, 1980). Puente and Bras (1987) argued that the proper error quantification of the model is even more important than the selection of the DA methods.

Different (subjective) methods have been proposed to calculate model error in case of KF. In Maybeck (1982), an overview of such methods is reported. One of the main issue of model error estimation is the required computational costs for large models. In addition, simple parameterization of the model error might be necessary due to the lack of available information (Dee, 1995). That is why, in most of the applications with KF, the common approach is to manually calibrate the model error (Verlaan, 1998). Another approach is to use least square method in order to minimise the difference between the computed and observed covariance of the residuals (Verlaan, 1998).

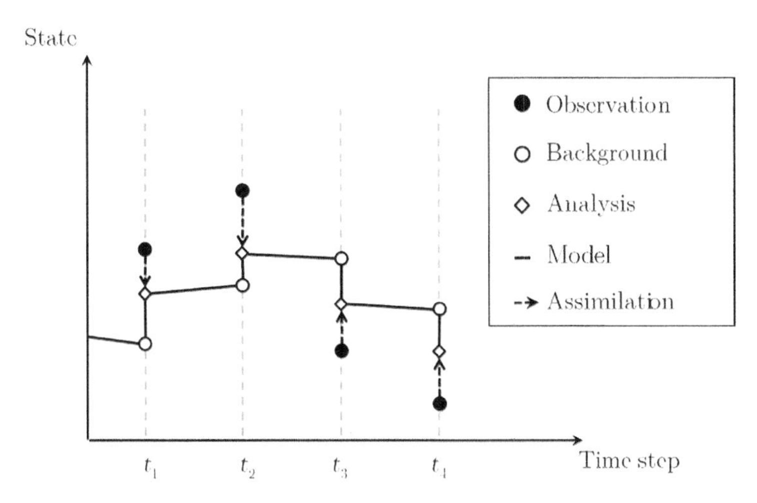

Figure 3.2. Schematic representation of the KF (adapted from Walker and Houser, 2005)

3.5 ENSEMBLE KALMAN FILTER

In the previous section it is seen that KF is optimal only in case of linear systems. However, in case of non-linear systems the extended Kalman filter (Kalman, 1960; Verlaan, 1998; Madsen and Cañizares, 1999; Aubert et al., 2003) or Ensemble Kalman filter (EnKF, Evensen, 2006) can be used to overcome the limitation of the linearity assumption. EnKF (Heemink et al., 2001; Reichle et al., 2002; Evensen, 2003; Weerts and El Serafy, 2006) is a widely used data assimilation method in hydrological applications. The main idea of the EnKF is to estimate the updated probability density function (pdf) of the model states, in an efficient recursive way, as a combination between data likelihood and forecasted pdf of model states by means of a Bayesian update. In EnKF the forecasted pdf estimate is represented with a set of random samples computed using a Monte Carlo method. EnKF can be divided in two steps, named the forecast and update (or analysis) steps. In the forecast step the forecasted matrix of the ensemble of model states is estimated as:

$$\mathbf{X}_t^- = \left(\mathbf{x}_{t,1}^-, \mathbf{x}_{t,2}^-, \cdots, \mathbf{x}_{t,i}^-, \cdots, \mathbf{x}_{t,N_{ens}}^-\right) \tag{3.12}$$

where $\mathbf{x}^-$ is the forecasted (or background) matrix of the model state for a given ensemble member i and N_{ens} is the total number of ensemble members. In Figure 3.3 three ensemble members $x^-_{t;1}$, $x^-_{t,2}$, and $x^-_{t,3}$ at time step t are represented with their probability density function (pdf) schematized using dashed dotted line. The ensemble mean of the forecasted state matrix:

$$\bar{\mathbf{x}}_t = \frac{1}{N_{ens}} \sum_{i}^{N_{ens}} \mathbf{x}_{t,i} \tag{3.13}$$

is used to derive the ensemble anomaly (Clark et al., 2008) for each ensemble member:

$$\mathbf{E}_t = \left(\mathbf{x}_{t,1}^- - \bar{\mathbf{x}}, \mathbf{x}_{t,2}^- - \bar{\mathbf{x}}, \cdots, \mathbf{x}_{t,i}^- - \bar{\mathbf{x}}, \cdots, \mathbf{x}_{t,N_{ens}}^- - \bar{\mathbf{x}}\right) \tag{3.14}$$

In this way, the evaluation of the model error covariance matrix is performed as proposed by Evensen (2003):

$$\mathbf{P}_t^- = \frac{1}{N_{ens} - 1} \mathbf{E}\mathbf{E}^{\mathrm{T}} \tag{3.15}$$

When an observation became available at the time step t a perturbed (with noise ν_t) normally distributed measurement vector $\mathbf{z}^o$ is generated. The product of $\mathbf{Hx}_{t,i}$

indicates the ensemble vector of model measurements. Each member of the measurement vector is assimilated with a member of the forecasted state matrix to generate an updated estimate of the model pdf. The update equation is

$$\mathbf{x}_{t,i}^{+} = \mathbf{x}_{t,i}^{-} + \mathbf{K}_t \cdot \left(z_{t,i}^{o} - \mathbf{H}\mathbf{x}_{t,i}^{-}\right) \tag{3.16}$$

where $\mathbf{K}_t$ estimated based on Eq.(3.8) and $\mathbf{x}^{+}$ is the update (or analysis) model state matrix. In Figure 3.3, the pdf of the posterior estimate (solid line), represented by $x^{+}_{t,1}$, $x^{+}_{t,2}$, and $x^{+}_{t,3}$, is obtained by combining $x^{-}_{t,1}$, $x^{-}_{t,2}$, and $x^{-}_{t,3}$ with the perturbed observations $z^{o}_{t,1}$, $z^{o}_{t,2}$ and $z^{o}_{t,3}$, which approximate the observation error pdf in dotted line. Errors of measurements and states are assumed to be uncorrelated (Neal et al., 2007).

It has been shown by various authors (e.g., Murphy, 1988; Anderson, 2001; Pauwels and De Lannoy, 2009) that the performance of ensemble forecast is influenced by the spread of the ensemble. In addition, considering that with the EnKF the model error is quantified as a function of the model realizations spread, the ensemble size has to be chosen carefully since it considerably influences the computational efficiency of the EnKF. For this reason it is important to perturb the system in a way to obtain a reliable spread of the ensemble within a meaningful range (De Lannoy et al., 2007).

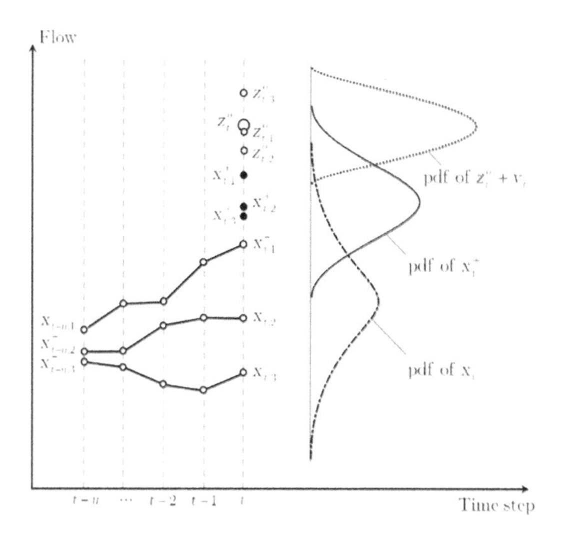

Figure 3.3. Schematic representation of the EnKF (adapted from Komma et al. 2008)

In order to estimate the value of ε_p and model realization *nens*, the approach proposed by Anderson (2001) to evaluate the quality of the ensemble spread is used. This author introduced the Normalized RMSE Ratio (NRR) estimated as:

$$NRR = \frac{R_a}{E[R_a]} \tag{3.17}$$

where R_a is the ratio between the time-averaged root mean square error (RMSE) of the ensemble mean R_1 and the time-averaged mean RMSE of the ensemble members R_2 (Anderson, 2001; Moradkhani et al., 2005a; Brocca et al., 2012).

$$R_a = \frac{R_1}{R_2} \tag{3.18}$$

$$R_1 = \frac{1}{T}\sum_{t=1}^{T}\sqrt{\left[\left(\frac{1}{N_{ens}}\sum_{i=1}^{N_{ens}} z_{t,i}\right) - z_{t,i}^{o}\right]^2} \tag{3.19}$$

$$R_2 = \frac{1}{N_{ens}}\sum_{i=1}^{N_{ens}}\sqrt{\frac{1}{T}\sum_{t=1}^{T}\left(z_{t,i} - z_{t,i}^{o}\right)^2} \tag{3.20}$$

where N_{ens} is the ensemble size, T is the simulation period. If the observed value z^o is statistically indistinguishable from the N_{ens} ensemble members, the expected value of R_a should be:

$$E[R_a] = \sqrt{\frac{N_{ens} + 1}{N_{ens}}} \tag{3.21}$$

If NRR > 1 indicates that the ensemble has too little spread, while NRR < 1 is an indication of an ensemble with too much spread. Ideal ensemble generation should produce a NRR value close to unity.

3.6 Asynchronous Ensemble Kalman Filter

Sakov et al. (2010) introduced the asynchronous ensemble Kalman filter (AEnKF), a generalization of the EnKF, which uses past observations over a time window, at once, in order to update model state at current time step, similarly to 4D-Var methods (Rakovec et al., 2015). However, AEnKF does not require any adjoin model like in case of 4D-Var. Sakov et al. (2010). Figure 3.4 shows the difference of the model updating procedure using EnKF and AEnKF. It can be seen that in case of EnKF the model is updated using only the observation at the current time step t, while in case of AEnKF also the past observations up to t-3 (W=3) are included in the assimilation process.

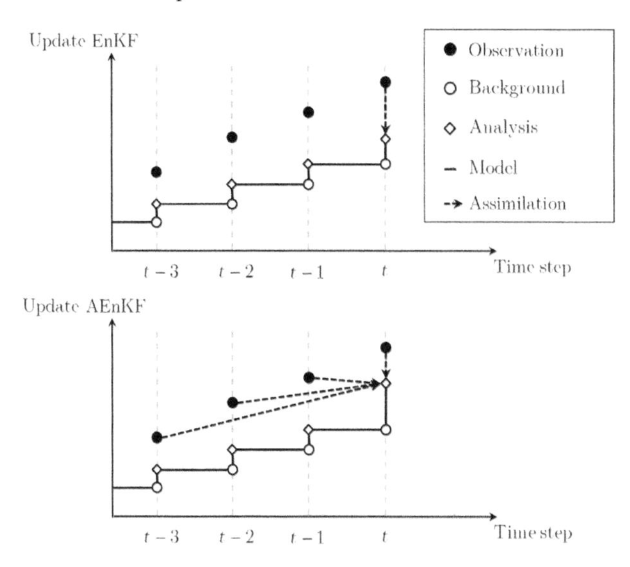

Figure 3.4. Difference between model updating procedure using EnKF and AEnKF in the simplified case of just one ensemble member and W=3

The structure of the AEnKF is based on the previous equations representing the EnKF. In particular, the model state matrix is augmented with the past forecasted observations from W previous time step. The new state augmented matrix (n_{state}+ W, 1) can be expressed as:

$$\tilde{\mathbf{x}}_t^- = \begin{bmatrix} x_{t,i}^- \\ h(x_{t-1,i}^-) \\ h(x_{t-2,i}^-) \\ \vdots \\ h(x_{t-W,i}^-) \end{bmatrix} \tag{3.22}$$

where $h(x_{t-W,i})$ is the model output corresponding to the time step t- W. In a similar way, with the new state definition, also the operator matrix $\mathbf{H}$ (n_{state}+ W, n_{state}+ W), observation covariance matrix $\mathbf{R}$ (n_{obs}+ W, n_{obs}+ W) and observation vector $\mathbf{z}$ (n_{obs}+ W, 1), can be expressed in their augmented form as:

$$\widetilde{\mathbf{H}}_t = \begin{bmatrix} H_t & & & & \\ & I_{t-1} & & 0 & \\ & & I_{t-2} & & \\ & 0 & & \ddots & \\ & & & & I_{k-W} \end{bmatrix} \tag{3.23}$$

$$\widetilde{\mathbf{R}}_t = \begin{bmatrix} \mathbf{R}_t & & & & \\ & \mathbf{R}_{t-1} & & 0 & \\ & & \mathbf{R}_{t-2} & & \\ & 0 & & \ddots & \\ & & & & \mathbf{R}_{t-W} \end{bmatrix} \tag{3.24}$$

$$\widetilde{\mathbf{z}}_t = \begin{bmatrix} \mathbf{z}_t \\ \mathbf{z}_{t-1} \\ \mathbf{z}_{t-2} \\ \vdots \\ \mathbf{z}_{t-W} \end{bmatrix} \tag{3.25}$$

Where I is the identity matrix having the same dimension of **H**, i.e. $(1,n_{state})$. It is worth noting that, even if the previous matrices are made augmenting past observations, Eq.(3.16) is solved only for the current time step t without updating past model states. Follow the Eq.(3.15) and Eq.(3.8) of the EnKF, also the matrices **P** and **K** change size in both rows and columns directions. In fact, an extra column in **K** corresponds to the gain due to one past observations. In case of W=0, the AEnKF formulation is identical to the one of EnKF. The characteristic of the AEnKF of adding past observations in order to improve the DA procedure at the current time step is very attractive in case of operational use thanks to the low calculation costs of such method.

4

ASSIMILATION OF SYNCHRONOUS DATA IN HYDROLOGICAL MODELS

This chapter describes the procedure developed to assimilate streamflow observations, synchronous in time, from StPh and StSc sensors within the semi-distributed model implemented in the Brue catchment, reported in Chapter 2. The influence of StPh sensor locations and varying observation accuracy (Experiment 4.1) on the assimilation of distributed streamflow observations is assessed. In addition, the usefulness of assimilating uncertain CS data, dynamic in space (Experiment 4.2) and intermittent in time (Experiment 4.3), derived from StSc sensors is demonstrated. A standard version of the Ensemble Kalman Filter is applied to a semi-distributed hydrological model of the Brue catchment. Realistic synthetic observations are used to represent distributed CS observations within the case study area.

This chapter is based on the following peer-reviewed journal publications:

Mazzoleni M., Alfonso L., Chacon-Hurtado J.C. and Solomatine D.P. (2015) Assimilating uncertain, dynamic and intermittent streamflow observations in hydrological models", Advances in Water Resources, 83, 323-339;

Mazzoleni M., Alfonso L. and Solomatine D.P. (2015) Effect of spatial distribution and quality of sensors on the assimilation of distributed streamflow observations in hydrological modelling, Hydrological Sciences Journal, accepted.

4.1 INTRODUCTION

In the correct operational forecasting practice, assimilation of real-time observations into hydrologic and hydraulic models is performed considering traditional, static sensors (StPh), which can be located at the outlet section of the catchment or distributed within it. In fact, spatially and temporally distributed measurements are needed in the model updating procedures due to the complex nature of the hydrological processes to ensure a proper flood prediction (Clark et al., 2008; Rakovec et al., 2012; Mazzoleni et al., 2015a). For this reason, it is important to assess the influence of StPh sensor locations on the results of DA procedures since different locations affect the hydrological model performance. Blöschl et al. (2008) proposed streamflow assimilation in grid-based operational flood forecasting systems; however they did not analyse the effect of varying distributed streamflow observations within the river basin. Recently, various authors assessed the effect of interior discharge gauges on hydrological forecasts (Xie and Zhang, 2010; Lee et al., 2011a; Chen et al., 2012; Rakovec et al., 2012; Lee et al., 2012; Mendoza et al., 2012; Mazzoleni et al., 2015a, 2016).

In addition to the impact of StPh sensor locations on DA performance, another issue is the correct evaluation of the uncertainty affecting the streamflow measurements (data quality). In such observations, errors can be related to an inappropriate water level (WL) measurement or to the wrong assessment of the rating curve used to transform values of WL into discharges (Clark et al., 2008). Di Baldassarre and Montanari (2009) proposed a procedure for quantifying uncertainty of streamflow data, with particular focus on the analysis of rating curve uncertainty, neglecting however the uncertainty in the water level measurements. They found out that the estimation of river discharge using the rating curve method is affected by an overall error, at the 95% confidence level, equal to 25.6% of the observed river discharge for the considered case study located in river Po, Italy. Usually, the uncertainty in streamflow measurements is often assumed to have either normal or lognormal distributions (e.g. Moradkhani et al., 2005b; Weerts and El Serafy, 2006; McMillan et al., 2013) coming from uncertain estimations of the rating curve. Clark et al. (2008) proposed a version of the EnKF by transforming observed and modelled streamflow to log space before computing the Kalman gain. Fowler and Jan Van Leeuwen (2013) investigate how the relaxation of the Gaussian assumption affects the observation impact (measured considering the sensitivity of the analysis to the observations, the mutual information, and the relative entropy) within the assimilation process. Although methods for DA and uncertainty analysis have evolved recently, studies on the influence of sensor locations and their accuracy on DA procedures and model performance are still limited whereas there is a need to research these issues deeper.

However, the implementation of a new network of StPh sensor can be expensive due to their initial and maintenance costs. For this reason, low-cost, mobile sensors and mobile communication devices (e.g. StSc sensors) can be a valid alternative to integrate existing network of physical sensors. Unfortunately, none of the previous flood-related studies deal with either specific StSc sensor networks or CS observatories: they consider neither the variable accuracy (uncertainty) of sensors within the basin dynamic at each time step, nor the intermittent nature of such observations. In oceanic and meteorological modelling, assimilation of distributed intermittent observations is often practised. Due to the irregular sampling times of oceanographic observations, most of the ocean data assimilation (ODA) systems use continuous approaches, as 3D-Var or 4D-Var methods, in order to assimilate these intermittent observations at their corresponding times. Huang et al. (2002) proposed an improved continuous data assimilation scheme in which an incremental analysis update strategy is combined with a continuous ODA model. In case of observations randomly distributed in time and space, Macpherson (1991) compared the repeated insertion (RI) method with an alternative intermittent analysis-forecast cycle (AF). Sinopoli et al. (2003) proposed a robust Kalman filtering formulation able to model the arrival of observations as a random process. In addition, c developed a discrete Kalman Filter which allows the assimilation of incomplete date series. Despite the approaches previously described, in this chapter we decided to use a more straightforward and pragmatic method, often used in real-time EWSs, similar to the approach proposed by Cipra and Romera (1997) in order to assimilate the intermittent observations into the hydrological model.

One main goal and innovation of this chapter is to assess the effect of StPh sensor locations and different observation accuracies on the assimilation of distributed streamflow observations into a semi-distributed hydrological. Although it is not intended to provide optimal layout of sensor locations, the results of this chapter can be used to draw new criteria for streamflow network design, and complementing recent studies in such research area (e.g. Alfonso et al., 2010; Kollat et al., 2011; Alfonso and Price, 2012) . The second goal of this chapter is to demonstrate the usefulness of assimilating uncertain CS streamflow observations, intermittent in time and space, from StSc in the context of semi-distributed hydrological modelling. In particular, a realistic representation of CS streamflow observation from StSc is considered in case of citizens participating in information capture along with (or instead of) using the traditional StPh stations.

4.2 METHODOLOGY

In this chapter, the observations from both StPh and StSc are considered synchronous, i.e. their arrival time matches the model time step. In order to assimilate synchronous streamflow observations, from StPh and StSc sensors, within the semi-distributed hydrological model structures implemented for in the Brue catchment (see section 2.2.2 page 32), an EnKF is used in each sub-catchment s. Two different flood event (A and B, described in section 2.2.2) are used to assess the performances of the DA method in case of assimilation of CS observations. Due to the fact that distributed real-time flow data, from StPh and StSc sensors, are not available at the time of this chapter, synthetic realistic observations, with variable uncertainty in time and space and intermittent behaviour, are calculated. Below, the description of the DA method setup, assimilation of intermittent observations and the estimation of realistic synthetic observations are reported.

4.2.1 Assimilation of intermittent observations

In case of streamflow observations, continuous in time, from StPh and StSc sensors, a standard version of the EnKF is used. Despite the approaches proposed in ODA, a more straightforward and pragmatic method, often used in real-time EWSs, similar to the approach proposed by Cipra and Romera (1997) is used in this chapter in order to assimilate the intermittent observations into the hydrological model. It is a standard assumption of DA methods that between two update steps might occur different forecast steps.

For this reason, in this section one of the common approaches used to account for the different updating frequency is described. With updating frequency we mean how often an observation becomes available and the consequent assimilation into the hydrological model. The idea is to update the model states matrix $\mathbf{x}$ when observations are available and then forecast the discharge at the next time step, as usually done in DA approaches. During the time steps with no observations, the model states are estimated and forecasted using the KMN model, while the forecasted value of the error covariance matrix $\mathbf{P}$ (needed to assess the Kalman gain matrix $\mathbf{K}$ at the time step t+1) is estimated using Eq.(3.15) as the function of the forecasted model states (see Figure 4.1).

This approach is similar to the one proposed by Cipra and Romera (1997) which assume that when the update step is missing, the state covariance error (and actually the complete state probability distribution) does not change at that time step.

$$\mathbf{P}_t^+ = \mathbf{P}_t^- \tag{4.1}$$

For example, if observations are coming intermittently every four hours, then the model states are updated regularly with the EnKF for four hours. Instead, for the consecutive next four hours, the model runs without filtering propagating the error covariance matrix in time for the consecutive next four hours.

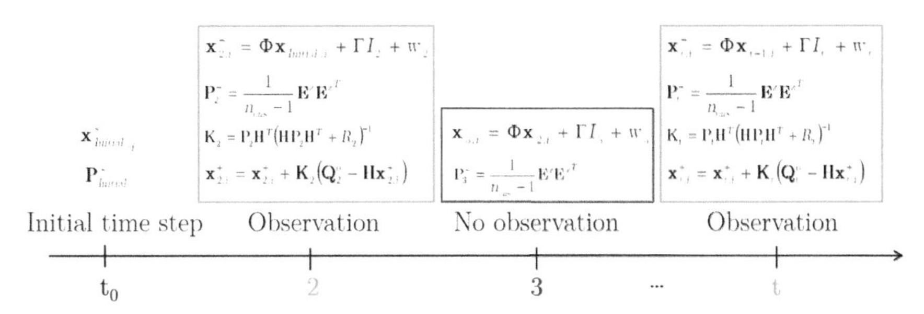

Figure 4.1. Graphical representation of the methodology used to assimilate intermittent observations in experiment 3 through EnKF algorithm.

This method can be applied in case of any updating frequency having minimum time step coincidence with the model time step. In most the hydrological application the main assumption is that the updating frequency is fixed. For example, in Rakovec et al. (2012) fixed the updating frequency as 6, 12 and 24h. However, in this chapter, the updating frequency it is not a priori decided but it depends to the random nature of dynamic and intermittent streamflow observations.

4.2.2 Observation and model error

In order to implement EnKF, an ensemble of model realizations is generated in to take into account the uncertainty related to the forcing inputs and model parameters. In this chapter the ensemble is generated perturbing the forcing data and the parameters for ach sub-catchment s as follows:

$$I_t'^s = I_t^s + U(-\varepsilon_I \cdot I_t^s, +\varepsilon_I \cdot I_t^s) \tag{4.2}$$

$$p'^s = p^s + U\left(-\varepsilon_p \cdot p^s, +\varepsilon_p \cdot p^s\right) \tag{4.3}$$

where I' is the ensemble of perturbed input data in each sub-catchment at time t, U is the uniform distribution, p' is the ensemble of perturbed parameter model p, ε_I is the fractional input error assumed equal to 0.2 (Clark et al., 2008; McMillan et al., 2013) and ε_p is the fractional parameter error. In this chapter, only the model parameter c (see Eq.(2.3)) is perturbed, keeping constant the number of reservoirs n in each sub-catchment. Eqs.(3.17), (3.18), (3.19), (3.20) and (3.21) are used in order to estimate the value of ε_p and model realization *nens* considering that a value of NRR greater than 1 indicates a small spread of the ensemble while value of NRR lower than 1 shows that the spread of the ensemble it is big. As a result of this analysis it is decided to set N_{ens}=65 and ε_p equal to 0.9 and 0.5, for the MS1 and MS2, which provided a value of NRR equal to 0.89.

An important issue is the proper quantification of the observation error as it is connected to the spread of the ensemble of synthetic observations and the consequent EnKF performance. On the one hand, observations having low error will provide a limited variability of $\mathbf{Q}^o$; on the other hand, a higher observation error will be reflected in a higher spread of the vector $\mathbf{Q}^o$. As mentioned by Clark et al. (2008) the standard deviation of the error σ_Q in streamflow measurements can be expressed as the linear function of Q^{true} at a given measurement location:

$$R_t = (\alpha_t \cdot Q_t^{true})^2 \tag{4.4}$$

where α is the coefficient of variation related to the uncertainty in the discharge measurement (described in the next section). The principle behind the Eq.(4.4) is that the high values of discharge are assumed to be more uncertain than the small values.

4.2.3 Generation of synthetic observations

An ensemble of synthetic streamflow observations $\mathbf{Q_{obs}}$, for each single sub-catchment s in which the EnKF is applied, normally distributed with the mean Q^{true} and standard deviation σ_Q are generated for each sub-catchment as follows:

$$\mathbf{Q}_t^o = Q_t^{true} \cdot \gamma + \mathbf{v}_t = Q_t^{true} \cdot \gamma + N(0, R_t) \tag{4.5}$$

where γ is a parameter that accounts for the uncertain estimation of the synthetic discharge (see next sections). The approach used to generate the synthetic values of Q^{true} is very similar to the one used by Weerts and El Serafy (2006). In such approach, the forcing I is perturbed by means of a time series normally distributed with zero mean and given standard deviation.

A similar approach, termed "observing system simulation experiment" (OSSE), is commonly used in meteorology to estimate synthetic "true" states and measurements by introducing random errors in the state and measurement equations (Arnold and Dey, 1986; Errico et al., 2013; Errico and Privé, 2014).

OSSEs have the advantage of making it possible to directly compare estimates to "true" states and they are often used for validating DA algorithms.

4.3 EXPERIMENTAL SETUP

4.3.1 Experiment 4.1: Streamflow data from static physical (StPh) sensors

In this section, the experiments performed to analyse the effects of a) different StPh sensor locations, and, b) accuracy of the observed data, on the assimilation of distributed streamflow observations in semi-distributed hydrological models are described. In such analyses it is assumed that for each sub-catchment only one measurement station is installed. The updating frequency is considered equal to one model time step, i.e. 1 hour.

Effect of StPh sensor spatial distribution

The correct evaluation of interior streamflow sensors position is fundamental to proper predict the flood hydrograph, which may lead to better decisions that reduce flood risk. For this reason, in this chapter we follow 3 steps (reported in Figure 4.2) to assess the effect of assimilation of distributed streamflow observations within hydrological modelling. It is worth noting that in this analysis it is assumed that the source of the observational errors is exclusively due to their transformation from water levels to discharge via a rating curve.

Step 1: A preliminary analysis is carried out in order to assess the performance of the models in case of assimilation of distributed discharge observations in a set of main locations of the catchment (called "spatial configurations SC"). In Figure 4.2, the green areas represent the sub-catchments in which it is assumed that the streamflow observations are assimilated. For example, in SC1 observations are assimilated in all the sub-catchments of the entire catchment, in SC2 only in the sub-catchments with Horton order 1, in SC8 in the main river reach (Horton order 3) and SC10 In the sub-catchments located close to the outlet cross-section of the catchment. Due to the fact that the hydrological model implemented in each sub-catchment is a conceptual lumped model, the streamflow observation is considered measured from only 1 hypothetical sensor at the outlet of the sub-catchment. The number of hypothetical StPh stations related to the number of sub-catchments considered in each spatial configuration is reported in Table 4.1.

Table 4.1. Number of hypothetical StPh sensors according to the different spatial configuration of measurements within the Brue catchment

Configuration	1	2	3	4	5	6	7	8	9	10
Number of StPh sensors	68	1	20	13	17	18	36	18	15	7

Step 2: In order to assess the responsiveness of additional measurements in a single location, streamflow observations are assumed to be available at any moment only in a single sub-catchment. The model accuracy is evaluated by the NSE index which compares the simulated and the observed discharge hydrograph at the catchment outlet. The single NSE values of each sub-catchment *s* are then normalized resulting in a normalized NSE (NNSE):

$$NNSE^{z} = \frac{NSE^{z}}{\max(NSE^{1}, NSE^{2}, \cdots, NSE^{z}, \cdots, NSE^{S})} \tag{4.6}$$

where *S* is the total number of sub-catchments The NNSE is used to compare the results obtained in case of different flood events since the magnitude of NSE can be different. Therefore, the NNSE are grouped in four classes for different flood events. In this way, the sub-catchments that induce a significant improvement in the flood hydrograph are identified for both MS1 and MS2 and a given flood event.

Step 3: Based on the class and the NNSE value of each sub-catchment for the two flood events, different scenarios of sensor locations that would give the best model improvement are introduced. The procedure used in this chapter to assess such scenarios is schematized as:

1. Two sensors, in the northern main river branch, located in the sub-catchments with highest class and with the corresponding highest NNSE;
2. Two sensors, in the southern main river branch, located in the sub-catchments with highest class and with the corresponding highest NNSE;
3. Two sensors, in the two opposite river branches, located in the sub-catchments with highest class and with corresponding highest NNSE ;
4. Sensors located in the sub-catchments considered in point 1 and 2;
5. Two additional sensors to those considered in point 4, located in the sub-catchments having high class but lower NNSE values than those in point 4, towards downstream. In this way, it is possible to assess the responsiveness of downstream sub-catchments in the streamflow assimilation;
6. Three additional sensors to those in point 4, located in random sub-catchments having high class but lower NNSE values than those in point 4. This scenario is included in the procedure in order to assess the influence of the total number of sensors;

It is worth noting that this procedure does not aim to exhaustively search for all N possible sensor location combinations, or replace traditional optimisation methods, but only those related to the classes that show model improvement at the outlet section of the catchment. In Table 4.2 the total number of sensors used in the different scenarios is reported.

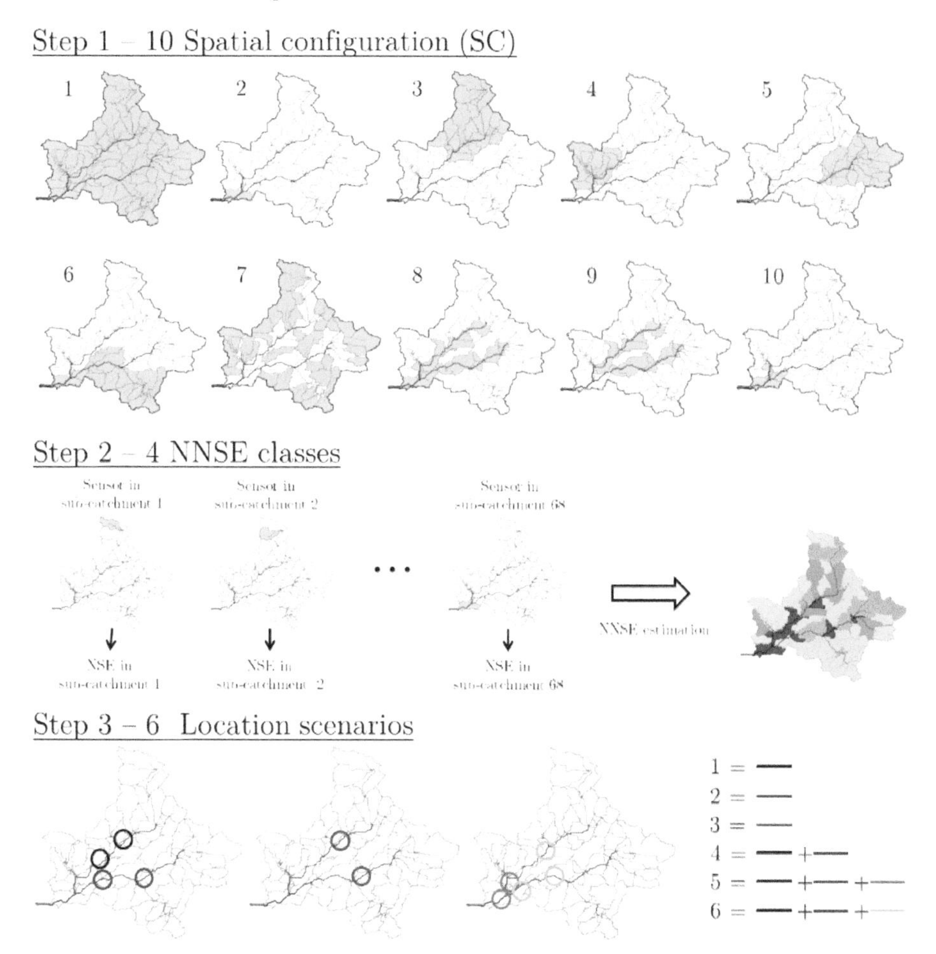

Figure 4.2. Representation of the 3-step method used to assess the effect of StPh sensor location on the DA performances. The NNSE values and location of sensors showed in step 2 and 3 are just hypothetical.

Table 4.2. Number of sensors according to the six scenarios of sensor locations within the Brue catchment.

Scenario	1	2	3	4	5	6
Number of sensors	2	2	2	4	5	7

Effect of data quality from StSy sensors

In this section, different sources of uncertainty which might affect the quality of the observed streamflow data are described. As pointed out by Clark et al. (2008), the uncertainty coming from discharge observations can be due to: a) incorrect estimation of water level at a given location, or b) inaccurate (uncertain) transformation of water level into discharge (rating curve). As reported by Di Baldassarre and Montanari (2009), the uncertainty induced by imperfect measurement of river stage can be negligible in case of using a static physical sensor. For this reason, in most of the data assimilation application in hydrology only the errors caused by an inaccurate rating curve are considered. However, we will consider this source of uncertainty in order to assess the effect on the DA procedure. Under these assumptions, Weerts and El Serafy (2006) and Clark et al. (2008) assumed that the error in the streamflow observations should be quantified as a noise terms normally distributed, with zero mean and given variance, described Eq.(4.4) with assumed to be equal to 0.1 (Weerts and El Serafy, 2006; Clark et al., 2008; Rakovec et al., 2012). In order to assess the effect of different observational errors on DA performance, different sources of uncertainty are considered, assuming a perfect forecast.

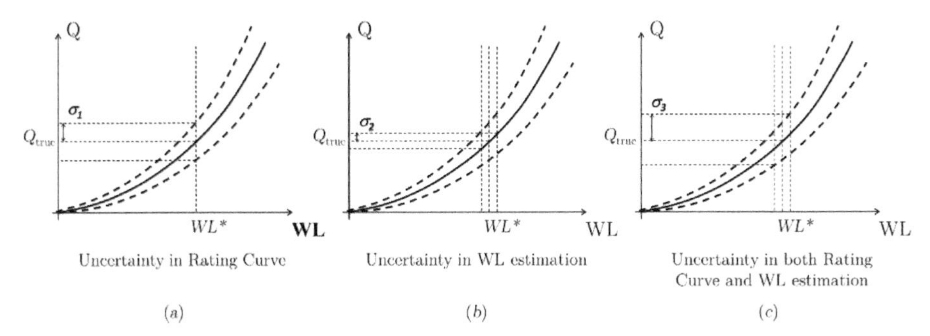

Figure 4.3. Representation of main source of uncertainty in streamflow data.

Figure 4.3 and Figure 4.4 illustrates various types of observational error; they can be characterised as follows:

- **Uncertain rating curve (ErrRC)**: uncertainty comes from an inadequate estimation of the rating curve used to transform water level into discharge, whereas the uncertainty induced by imperfect measurements of water level is assumed to be negligible. Normal distribution of the observational error is assumed, with a value of α_{RC} equal to 0.1 as usually proposed in hydrological data assimilation applications (e.g. Weerts and El Serafy, 2006; Clark et al.,

2008). This type of observational error will be used to assess the effect of sensor distribution and propose the possible network of static physical sensors.

- **Uncertain WL estimation (ErrWL)**: uncertainty comes from the imperfect measurements of water level, and not from the rating curve estimation. In this case, two different probability distributions, namely normal (WL1) and uniform (WL2), are used to characterize the vector of observations. It is worth noting that EnKF provides optimal results only if the distribution of the measurement vector is Gaussian (however can be used with any distribution). For these reason, the results that will be obtained in WL2 case can be considered as "sub-optimal" with respect to the EnKF assumptions. In addition, due to the fact that errors in the measurements are negligible (Di Baldassarre and Montanari, 2009) and smaller than the error in the rating curve estimation, the coefficient α_{WL} is assumed equal to 0.02.
- **Uncertain static sensors (ErrRC+WL):** the errors coming from uncertain rating curve and uncertain WL measurement are considered both. In this case, the value of the coefficient α in Eq (2) is assumed equal to the sum of the two previous coefficients α_{RC} (ErrRC) and α_{WL} (ErrWL1), resulting in an observational error normally distributed with zero mean and standard deviation equal to $\alpha_T \cdot Q_{true}$ with α_T set to 0.12.
- **Uncertain estimation of the synthetic discharge (ErrSD)**: In this case it is assumed that the value Q_{true} might be biased and different to the real one. For this reason, the parameter α (usually set to 1) is considered as a random uniform number between -0.3 and +0.3 in order to account the uncertainty in estimation of Q_{true}.

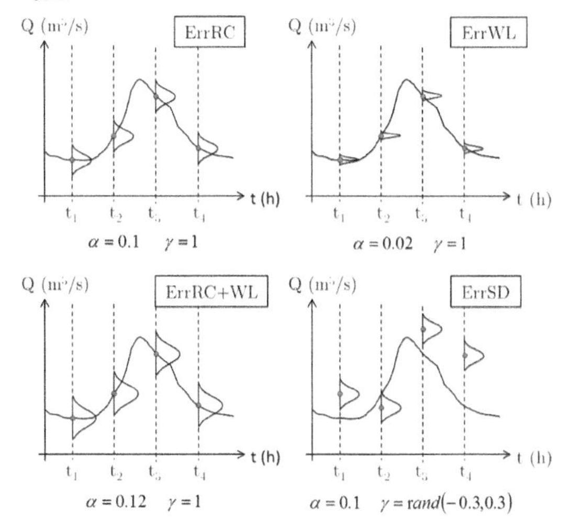

Figure 4.4. Values of α and γ according to the different types of uncertainty.

4.3.2 Experiment 4.2: Streamflow data from static social (StSc) sensors

The effect of the assimilation of observations, continuous in time and randomly distributed in the catchment at each time step from StSc sensors, on the outflow hydrograph is assessed for the two different model structures. As described in the previous section, in case of assimilation of streamflow observations from StPh sensors, the observational error is related to the uncertain estimation of the rating curve, and in particular, the standard deviation of such error is assumed to fixed in time and space (e.g. Clark et al., 2008; Rakovec et al., 2012; Weerts and El Serafy, 2006). However, in case of StSc sensors, uncertainty is manifested in a more complex manner. In addition to the poor quality of the rating curve, the incorrect measurements of WL according to the accuracy of different types of StSc sensors, is assumed to be the main sources of uncertainty. In fact, it is assumed that StSc are used by citizens which can move from one sensor to another one. For this reason, in order to represent the dynamic behaviour of citizens using StSc sensors from one sub-catchment to another one, the observational error R, is assumed to be function of time t and of the random distribution of the discharge at each time step in each sub-catchment z.

$$R_{z,t} = \left(\alpha_{z,t} \cdot Q_{z,t}^{true}\right)^2 \tag{4.7}$$

Consequently, the parameter α is assumed to be a stochastic variable, changing in time and space, inducing the resulting (see Figure 4.5).

No specific spatial sensor trajectory of the citizen moving from one StSc sensor to another one is considered in this chapter since this would require the introduction of assumptions about citizens' behaviour in case of flood event. The author is aware that this component would be extremely important in case of dynamic sensors but this could not be included in this chapter due to the lack of information about citizen engagement in monitoring river water level in the Brue catchment. This approach to describing the complex nature of uncertainty due to the presence of dynamic citizen moving across StSc sensors is one of the novelties of this paper.

For this reason, α is considered to be a random stochastic variable uniformly distributed in time t and space (sub-catchment s) as $U(\alpha_{min},\alpha_{max})$. α_{min} and α_{max} are set to 0.1 and 0.3 respectively to account for the low and high observational noise. In order to assess the influence of the variable discharge uncertainty within each spatial configuration of measurements sites, five different situations of spatial and temporal evolution of the dynamic sensors, during the flood events A and B, are randomly generated. Also in this particular experiment the discharge observations are assumed to be available at hourly time steps, i.e. an updating frequency equal to 1 hour. It is important to note that the discharge and states estimated at the time step t influence the discharge at the downstream sub-catchment in the time step $t+\Delta t$ since the time delay Δt is assumed in the flood propagation module.

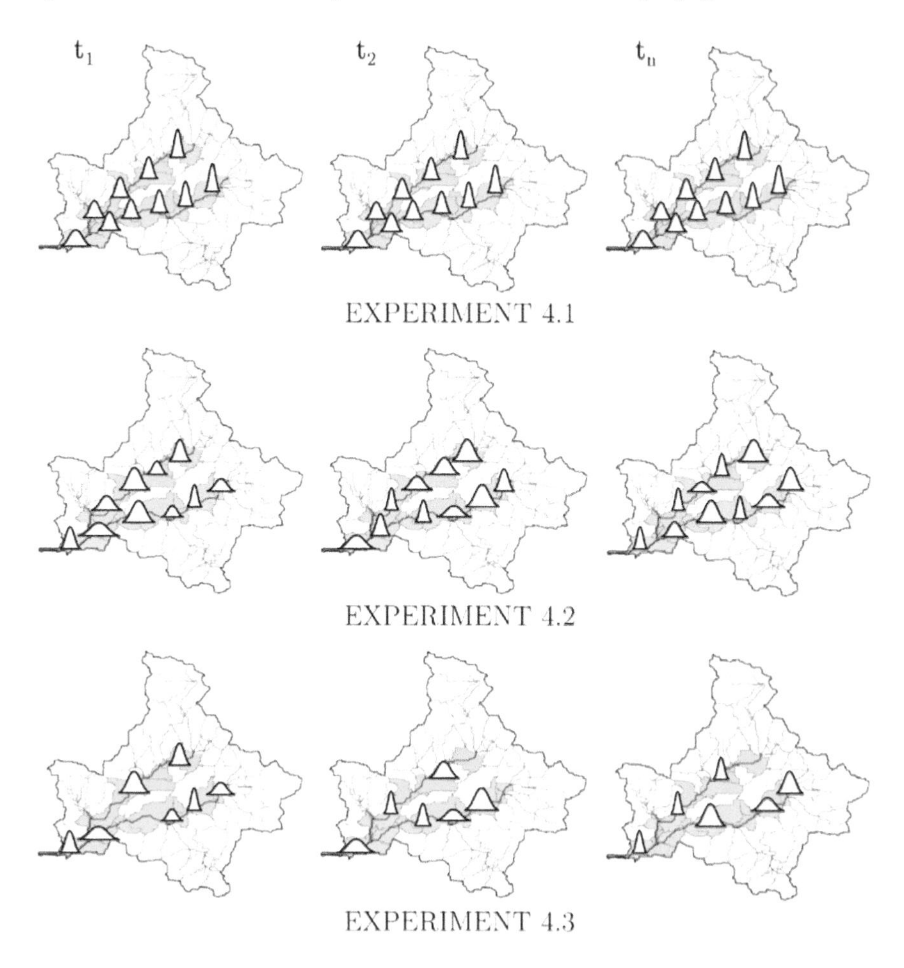

Figure 4.5. Qualitative representation of synthetic data noise $N\left(0, \alpha_{t,s} \cdot Q_{t,s}^{true}\right)$ *used in experiments 4.1, 4.2 and 4.3 in case of spatial configuration of sensors location 8 at different time steps* t_1, t_2 *and* t_n.

4.3.3 Experiment 4.3: Intermittent streamflow data from static social (StSc) sensors

As mentioned above, one of the objectives of this research is to understand the effects of the assimilation of intermittent observations on the outflow hydrograph generated by the two different model structures. The difference between experiments 4.2 and 4.3 is that in the first one, streamflow observations are considered continuous in time, while in experiment 4.3 these observations are intermittent. In case of StSc sensors it can happen that at a given location the discharge values might be measured only at some irregular time steps, as opposed to the continuous observations from traditional static stations. Intermittent sensors are a category of StSc sensors in which the observations is provided occasionally in time and space. As mentioned before, an example can be the picture of a river water level provided the camera (sensor) of a mobile phone. In this case the observation is carried out at one time step and at one particular location since the owner of the mobile phone might move and provide another water level observation in another location and different time step. In order to represent the intermittent nature of such observations, a random variable $\psi(t)$, which takes a random value of either 0 or 1 at different time steps, is introduced. In this way, the DA procedure is performed when $\psi(t)$=1 while in the opposite case no observation is available from the sensor and no assimilation is carried out. To summarize this concept, Eq.(4.4) is updated as:

$$\mathbf{Q}_t^o = \begin{cases} Q_t^{true} + N(0, R_t), & \text{if } \Psi_t = 1 \\ N.A., & \text{if } \Psi_t = 0 \end{cases} \tag{4.8}$$

In addition, this experiment considers five different cases of randomly generated intermittent signals of the sensors during both flood events, with the purpose of assessing the influence of the intermittent nature of the distributed observations of discharge.

4.3.4 Experiment 4.4: Heterogeneous network of static physical (StPh) and static social (StSc) sensors

In the previous sections, hypothetical randomized locations of StPh and StSc sensors are assumed. However, location of sensors should typically follow some rules and be subject to constraints. For example, existence of multiple StPh sensors along the main river reach is quite unlikely due to economical and management reasons. This experiment aims at assessing the influence of observations from heterogeneous sensors, StPh and StSc, in case of a realistic configuration of the latter within the basin.

Therefore, we focus on a realistic situation in which synthetic CS observations of WL are provided by means of dynamic sensors for both flood events A and B used in the validation phase (see Chapter 2). An example of CS observations, as previously described, might be the water level measurements taken with a StSc sensors made by a staff gauge (the reference sensor) and Quick Response (QR) codes. These observations will complement the existing staff gauges. For obvious reasons, the citizens' observations introduce additional uncertainty which is manifested by the random variable α used in Eq.(4.4). In this chapter, realistic citizen-based observations are emulated by the synthetic dynamic observations of discharge previously estimated in Experiment 4.2 and 4.3.

The location of staff gauges, within the Brue catchment, is selected with respect to two aspects: (a) observations along the main river reach provide significant model improvement and (b) the higher number of observations is expected to come from the more populated areas (see Figure 4.6).

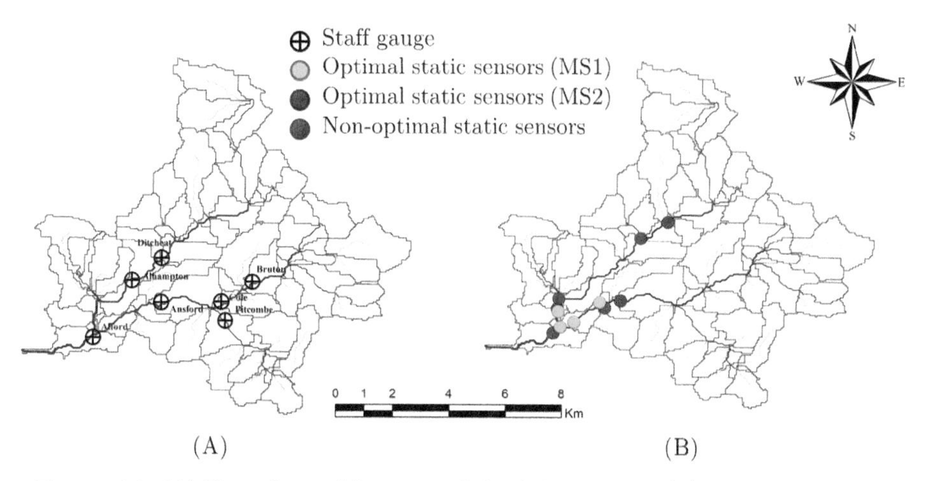

Figure 4.6. (A) Hypothetical location of the StSc sensors; (B) Optimal and non-optimal network of StPh sensors for MS1 and MS2 respectively.

Temporal variability in observations' frequency is taken into account as well: in a real application observations would arrive mostly at some specific moments during the day. For this reason, citizen-based observations are assumed to be available only between 9am and 5pm (Daylight hours in Table 4.3). In addition to the citizens' measurements, the observations coming from the 'trained volunteers' or 'highly engaged citizens' are assimilated as well. These observations are assumed to be concentrate around the peak discharge, continuously in time.

In this experiment, the synthetic observations of discharge coming from the few StPh sensors, which form an optimal and a generic non-optimal network of sensors (in terms of position and number of sensors, Figure 4.6), are combined with the observations from StSc sensors. These combinations result in 14 different settings described in Table 4.3.

Table 4.3. Description of the different settings introduced in experiment 4.4

	StSc sensors			StPh sensors	
Setting	**Intermittent**	**Daylight hours**	**Day and peak hours**	**Optimal**	**Non-Optimal**
1	-	X	-	-	-
2	X	X	-	-	-
3	-	-	X	-	-
4	X	-	X	-	-
5	-	-	-	X	-
6	-	X	-	X	-
7	X	X	-	X	-
8	-	-	X	X	-
9	X	-	X	X	-
10	-	-	-	-	X
11	-	X	-	-	X
12	X	X	-	-	X
13	-	-	X	-	X
14	X	-	X	-	X

The optimal configuration of sensors (i.e. sensors locations and the total number of sensors) is identified by solving an optimization problem by maximizing the NSE value for each one of the considered 6 location scenarios for StPh sensors presents in section 4.3.1. The optimal scenario of sensor locations is obtained as a compromise between total number of sensors and model improvement at the outlet section of the basin.

4.4 Results and discussion

This section reports the results of the four assimilation experiments previously described: 1) distributed StPh sensors; 2) StSc sensors; 3) intermittent StSc sensors; and 4) heterogeneous network of StPh and StSc sensors. Model performances without EnKF are reported in Figure 4.7 in case of MS1 and MS2 during the flood events A and B. As described previously, the poor performances of the hydrological model during flood event A can be due to the conceptual nature of the model itself or to an underestimation of the average precipitation used as input in each sub-catchment.

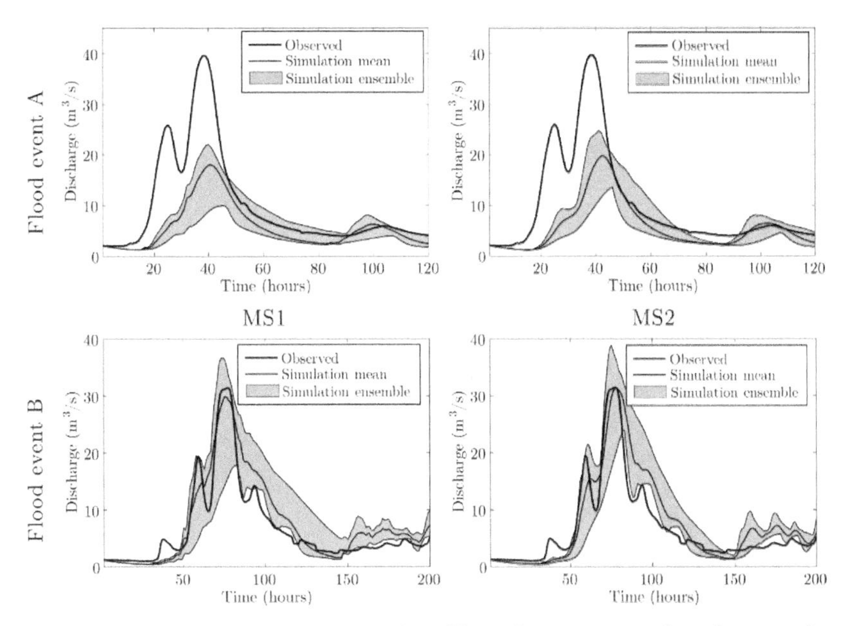

Figure 4.7. Streamflow estimation with 95% prediction interval without update in case of the MS1 and MS2 during flood event A and B.

4.4.1 Experiment 4.1

Effect of StPh sensors positioning

The main results obtained from the experiment 4.1 are summarized in Figure 4.8 and Figure 4.9. It is worth noting that the analyses described in this section are carried out considering only uncertain in rating curve (ErrRC), i.e. equal to 0.1. In Figure 4.8, the simulated ensemble mean estimated using EnKF in case of assimilation of streamflow observations in different spatial configurations (SC), 1, 2, 8, 9 and 10, are compared with the observed value of discharge. In particular, MS1 tends to provide higher model improvements than MS2 (e.g. flood event C and D). It can be seen that the spatial configuration which provides the best model performance in terms of NSE, for both the flood events, is the SC 1 (observations available in all sub-catchments). In the same way, SC 8 (observations from the main river reach with Horton order larger than three), provides a significant improvement to the model. Similar results are obtained by Rakovec et al. (2012) in case of distributed hydrological models. The comparable model performance, obtained for SC 1 and SC 8, can be explained by the fact that observations from river branches having order 1 (upstream part of the catchment) seem not to improve the model forecast. Assimilation of observations in SC 2, which can be seen as a standard situation in which the StPh sensors are installed in the outlet section of the catchment, does not provide any additional improvement to the model in case of uncertain observations. It is demonstrated that assimilation of observations coming from sub-catchments of Horton order 1 (SC 7) do not provide any improvement to the accuracy of the outflow hydrograph.

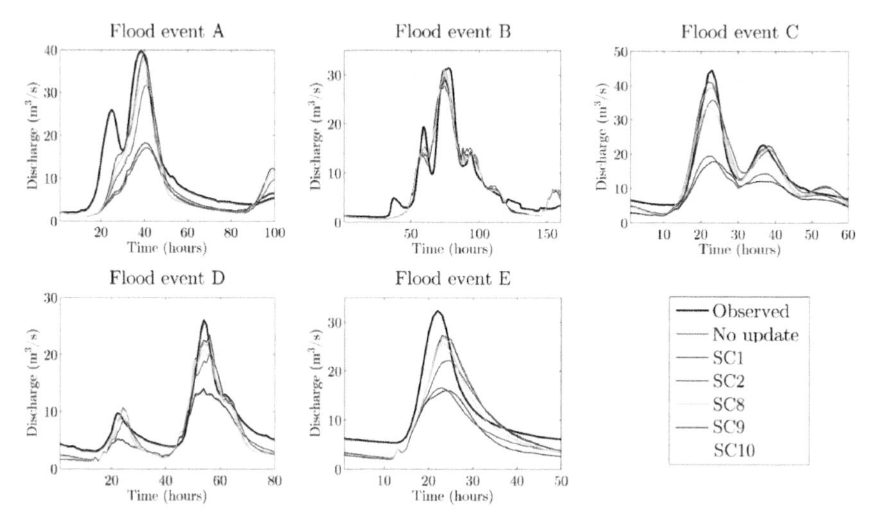

Figure 4.8. Comparison between observed hydrograph, model results and data assimilation results considering different sensor locations within main basin groups during all flood events in case of MS1.

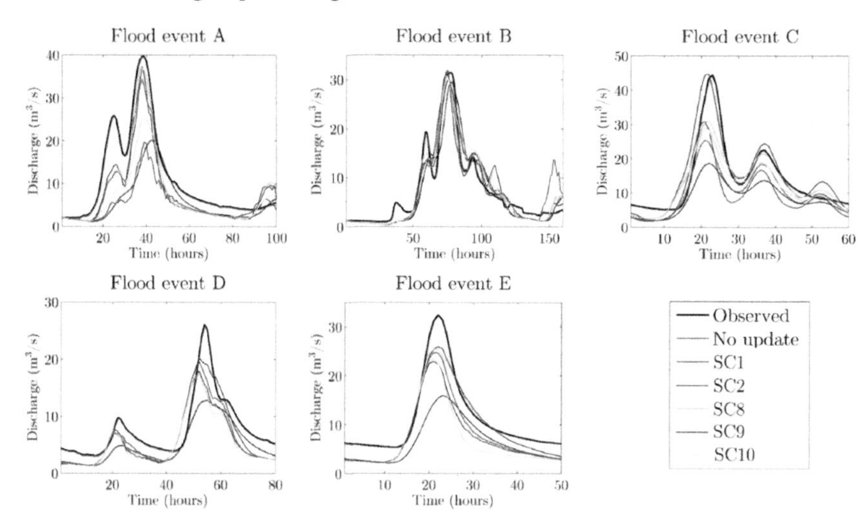

Figure 4.9. Comparison between observed hydrograph, model results and data assimilation results considering different sensor locations within main basin groups during all flood events in case of MS2.

Additional analyses are reported in Figure 4.10 during event A, which shows the error between mean of the true state and mean of the simulated state variable (water storage x) for two different sub-catchments in case of assimilation of static observations in SC 8, 9 and 10. The results in Figure 4.10 pointed out how, in case of MS1 during flood event A, simulated state variable tends to provide lower results than true states in all the considered spatial configurations. However, this error is lower in case of SC 8 in accordance with the results previously showed. These considerations are valid for both the sub-catchments analysed in this chapter.

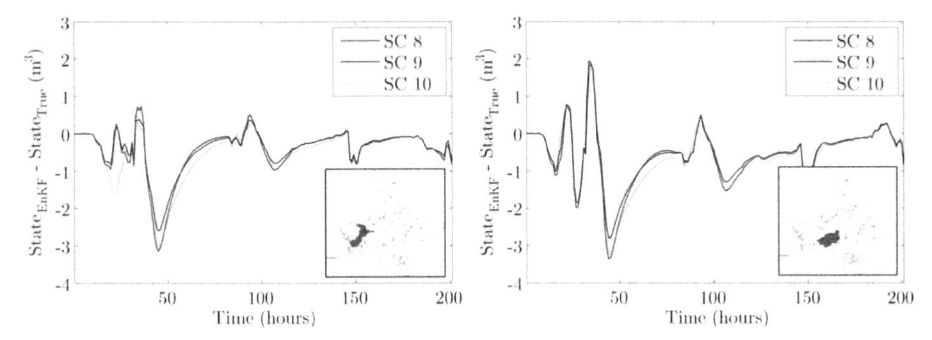

Figure 4.10. Error between simulated and synthetic true states **x** *(water storage) in case of 2 given sub-catchment assimilating streamflow observation in SC 8 , SC 9 and SC 10 during flood event A considering MS1.*

Based on previous results it can be seen that MS2 seems to be influenced by both the total number of sensors and their locations. A significant improvement of the model efficiency is achieved considering an appropriate compromise between position and number of sensors (SC 9) rather than focus only on a high number of sensors (SC 7) or on choosing the specific locations (SC 10). On the other hand, MS1 is not affected by the total number of sensors but only by their locations (SC 10). Similar results are obtained in case of flood event B.

At this point, it is possible to map the three different zones of sensor locations for both MS1 and MS2, which lead to quite different impacts of the model results (see Figure 4.11). Zone C corresponds to the observations which do not affect the outflow hydrograph. In general, discharge measurements located in the river channel with Horton order equal or bigger than 3 tend to better represent the outflow discharge. In particular, observations coming from Zone A, which corresponds to SC 10, lead to the best improvement of model output for MS1. On the other hand, sub-catchments within zone A are located in the downstream part of the catchment which is very often an urbanized area for which timely flood warnings could be especially important. The observations of discharge from this zone might indeed improve the model results in terms of outflow, but at the same time they would not provide enough time to react in the downstream areas where flood could be very damaging. On the other hand, assimilation of observations from Zone B (SC 9) might induce a non-substantial improvement in model accuracy if compared to SC 10, but it would give enough warning time before the estimated (high) flow reaches the downstream. Opposite results to the ones achieved with MS are obtained in case of MS2.

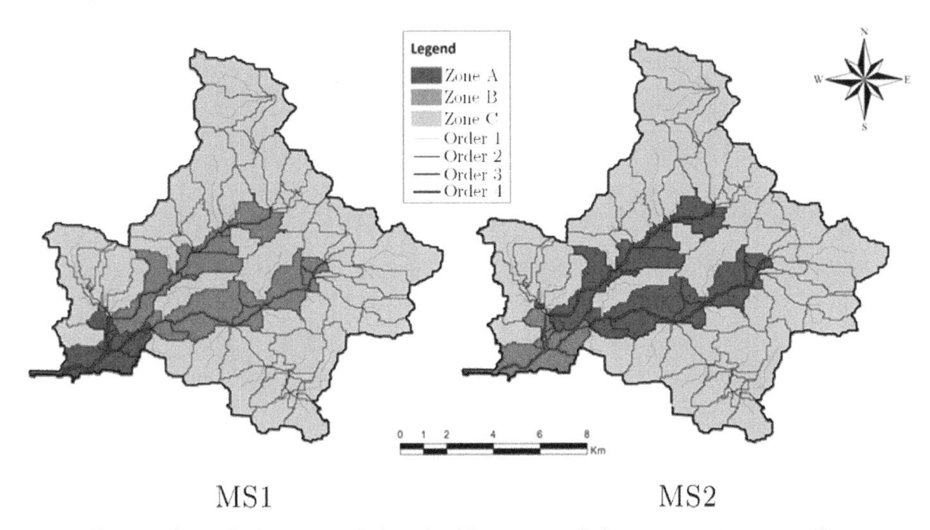

Figure 4.11. Indication of the ideal location of the uncertain streamflow observations to assimilate according to the improvement induced into the two proposed model structures.

The next step is to assess the DA performances considering available streamflow observations in given sub-catchments. Figure 4.12 shows the NNSE, which is used only to estimate the different model improvement classes, obtained from the assimilation of streamflow observations in a single sub-catchment in case of the two different model structures during the five flood events. The results show that location of the sub-basins which provides high NNSE values changes from MS1 to MS2. As previously showed, these sub-basins are mainly located along the main river channel (Horton order bigger than 3). However, such locations, for a given model structure, are very similar changing the type of the flood event. Overall MS2 provides lower NNSE values than MS1. In case of MS2 few sub-basins gives high NNSE values, while in MS1 high NNSE values are spread over a larger number of sub-basins.

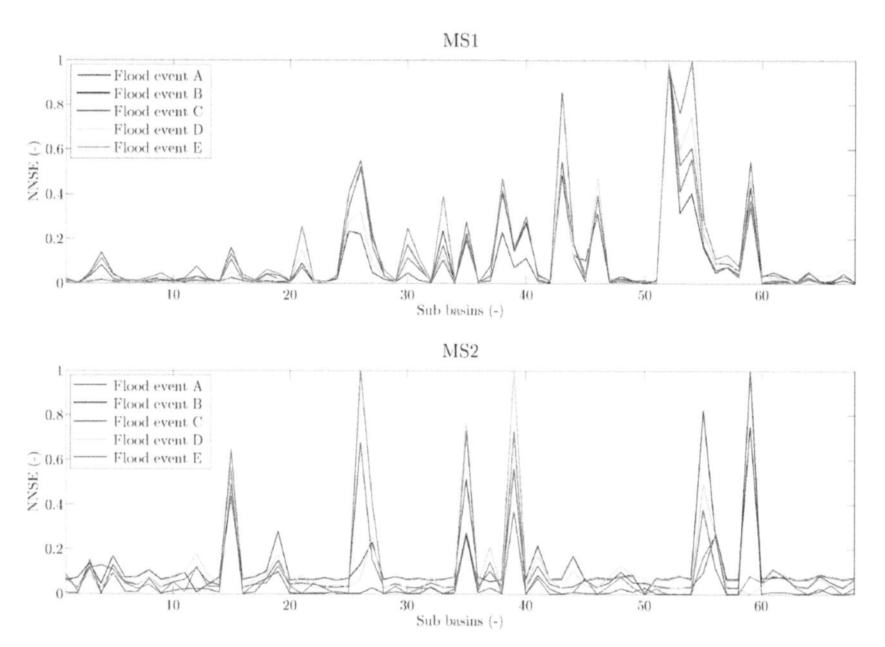

Figure 4.12. Indication of the responsiveness of each sub-basin, in terms of NNSE, according to the locations of the uncertain streamflow observations considering the two proposed model structures during the five flood events

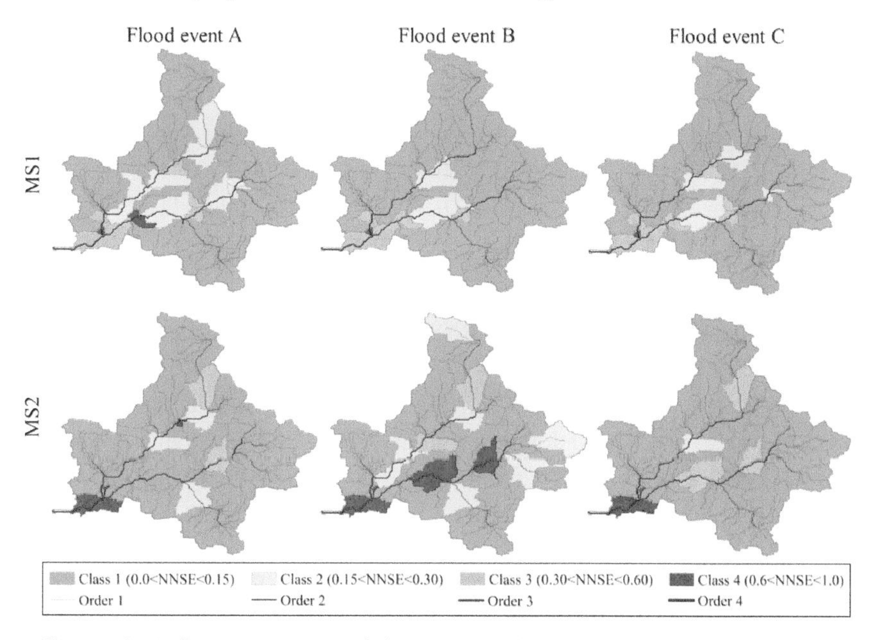

Figure 4.13. Representation of the estimated classes of each sub-catchment considering the two proposed model structures during the events A, B and C

Figure 4.13 shows the four classes obtained for flood events A, B and C in case of MS1 and MS2. Similar results are achieved for flood events D and E as well. As it can be seen, analogous location of the sub-basins having classes 3 and 4 are obtained for the three considered flood events for each structure.

Knowing the NNSE values in each sub-catchment, it is possible to assess the six sensor locations scenarios for both MS1 and MS2 during four flood events (see Figure 4.14).

Overall, sensor location do not drastically vary with the different flood events. For example, in case of MS1 and scenarios 3 and 4 the sensors are located in the same sub-basins (represented with light blue color) during the four analyzed events. Interestingly, in case of MS1 the sensors are located in the downstream part of the basin, which corresponds to the group 5 or Zone A previously analyzed. In case of MS2, sensors are mainly located in the upstream part of the Brue basin, as previously showed in the Zone A of such model structure.

Figure 4.15 shows the Taylor diagram for the five different events, six scenarios and two model structures. Taylor diagrams graphically summarize similarities between simulations and observations expressed in terms of root mean square error (RMSD), correlation and standard deviation. The closest is the simulation results to the observations (black cross) the better. The simulation results in case of flood event A are very closed between each other due to the good model performances without updates. As expected, the best model improvement is achieved for high number of sensors (i.e. scenario 6) for all the flood events. However, similar good model improvements can be also observed with scenarios 4 and 5. In case of flood events B and E, MS2 provides better model results than MS1. On the other hand, in case of flood events C and D, MS1 outperform MS2. Good correlation values are achieved in all flood events. The low standard deviation values are due to the underestimation of the simulated discharge values without model update.

Figure 4.16 shows the relative NSE (RNSE) expressed as difference between the NSE values of each given scenario and the one of scenario 6, i.e. the one that provides best model results, during the five flood events. Overall, in both structures, high RNSE values are achieved with scenarios 1, 2 and 3. Such values decrease at scenario 4 and 5. The main differences between MS1 and MS2 is that, in average, the RNSE values of scenario 4 are higher in MS2 than MS1. This means that MS2 is more sensitive to the total number of sensors than MS1. Low RNSE improvement are showed in case of event A due to the already high NSE value without model update.

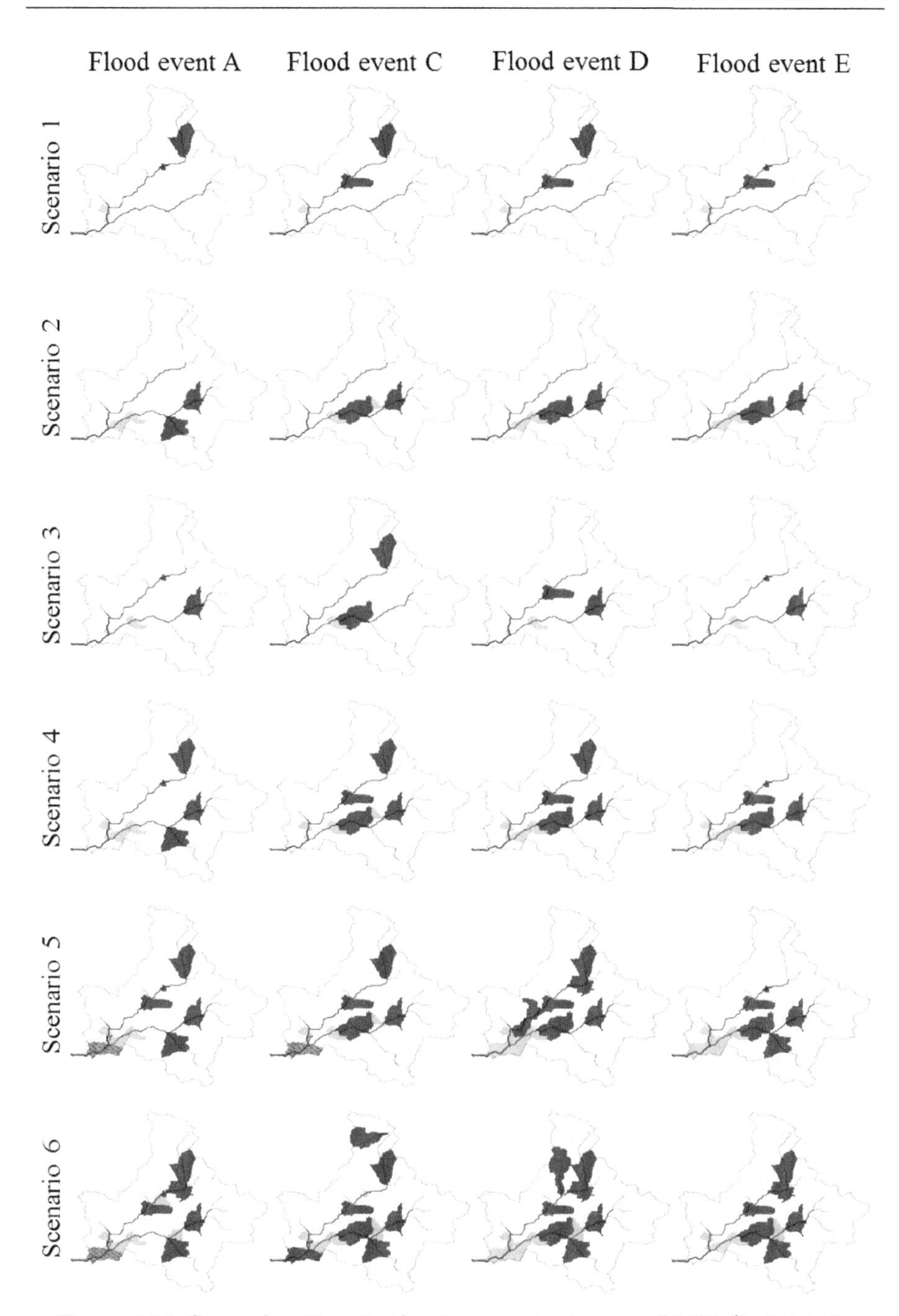

Figure 4.14. Sensor locations in the six scenarios in case of MS1 (in light blue colors), MS2 (in magenta color) during events B, C, D and E. The dark blue areas indicates the sensor location equal for both MS1 and MS2.

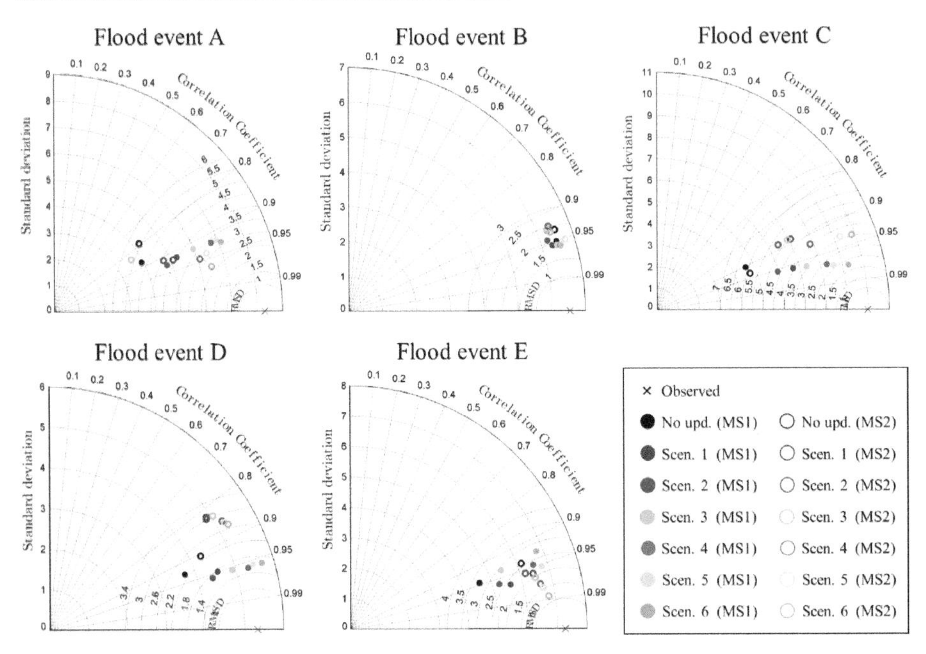

Figure 4.15. Taylor diagrams comparing observation with simulations obtained with MS1 and MS2 during all flood events in case of different sensor location scenarios.

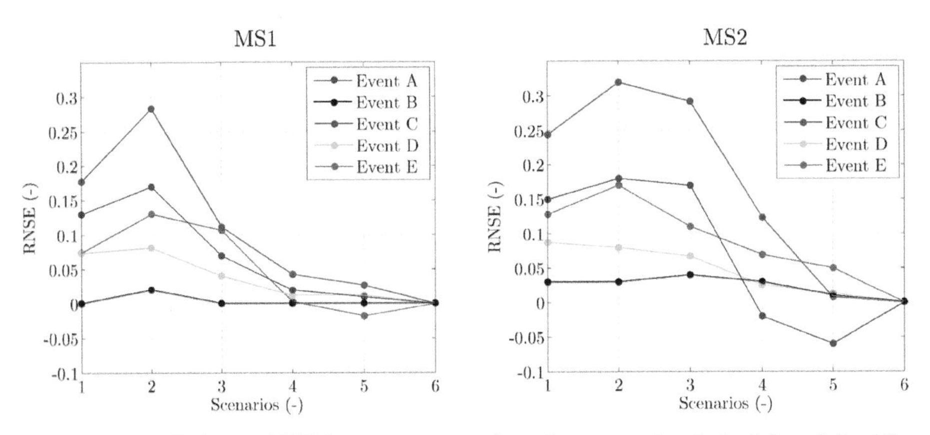

Figure 4.16. Relative NSE between sensor location scenarios 1, 2, 3 4 and 5 with scenario 6 for MS1 and MS2 during all flood events.

Figure 4.17 shows the relation between NSE and lead time values in case of MS1 and MS2 during flood events C, D and E. It can be observed that MS2 tends to the NSE values without model update faster (after 4hours) than MS1. In all flood events, scenario 6 is the one which provides highest model performances. It can also be seen that scenario 4 gives similar model results than scenario 6 in case of MS1 for different lead time values, as previously demonstrated. Lowest performances are obtained with scenarios 1, 2 and 3.

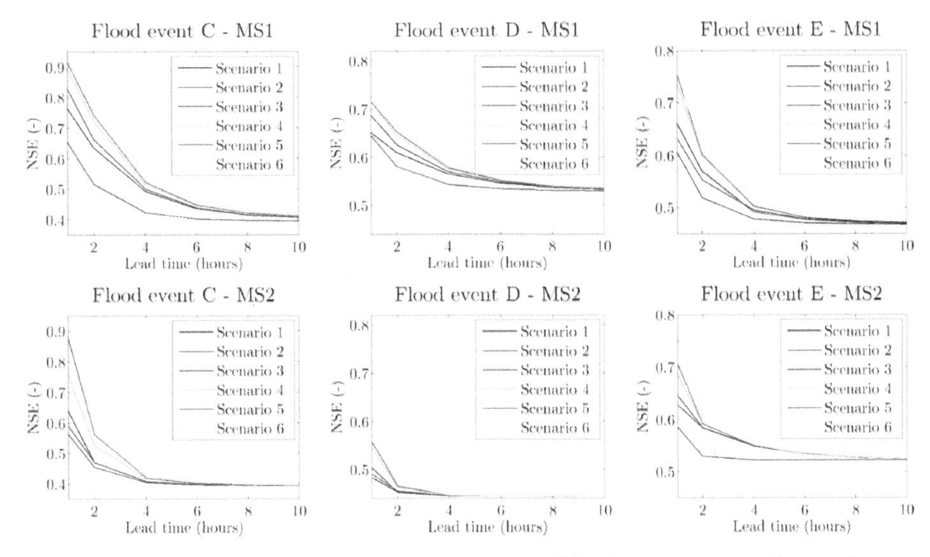

Figure 4.17. NSE obtained for different values of lead time in case of assimilation of streamflow observations in the six sensor locations scenarios during flood events C, D and E.

Effect of observation accuracy from StPh sensors

In this analysis, the effects of different observation accuracies on the model performances are assessed. Firstly, we considered the assimilation of streamflow observation within the main Brue Basin groups in case of only MS1, and then within scenario 6 for both MS1 and MS2 during the flood events C, D and E.

From Figure 4.18 it can be seen that the best model improvement is achieved assuming the error coming from the measurements only, i.e. ErrWL1 with normal or ErrWL2 with uniform distribution, in all the considered flood events in case of sensors located in the sub-basins of group 3. It can be noticed that better results are obtained in case of uncertain biased streamflow values (ErrSD) than considering only ErrRC. Overall, the smallest model improvements are achieved considering ErrRC+WL. An important results is that MS2 seems to be more sensitive to the proper definition of observational error than MS1.

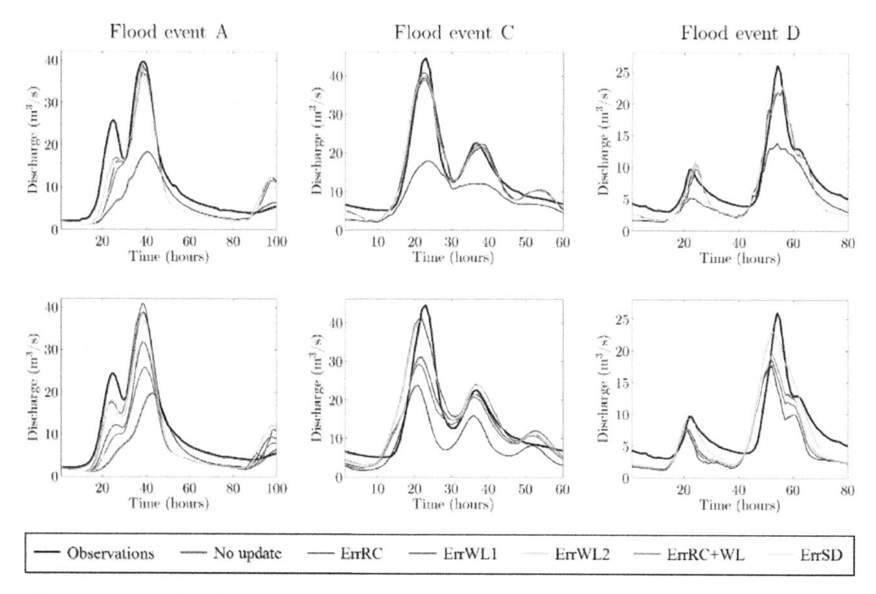

Figure 4.18. Outflow hydrographs obtained in case of four group 3 of spatial sensor locations in case of MS1 (first row) and MS2 (second row) considering different type of observational error in the DA procedure

Figure 4.19 shows the NSE values for MS1 and MS2, in case of different observational errors definition during events B, C, D and E. Sensors are assumed located in groups 3, 4 and 5. In case of events C and D, MS1 gives higher NSE values than MS2. As previously demonstrated, the variability of NSE values is higher in MS2 than MS1 for different observational errors. In addition, small differences between the NSE values are obtained between with ErrWL1 and ErrWL2.

Figure 4.20 shows the prediction of flood events C, D and E in case of sensor located according to scenario 6 in case of MS1 and MS2. The results obtained for shorter lead times are in agreement with the results shown in Figure 4.19 achieved for different sensor locations. Overall, MS1 seems to be less sensitive to the observational errors than MS2 also for high lead time values. However, large variability of NSE is achieved during event E using MS1. The best predictive efficiency is obtained for the both model structures in case of ErrWL, as previously described in Figure 4.18 and Figure 4.19 in case of lead time of 1 hour. In average, ErrRC+WL provides the lowest NSE values. Small difference between NSE obtained in case of ErrRC and ErrSD is showed for both MS1 and MS2.

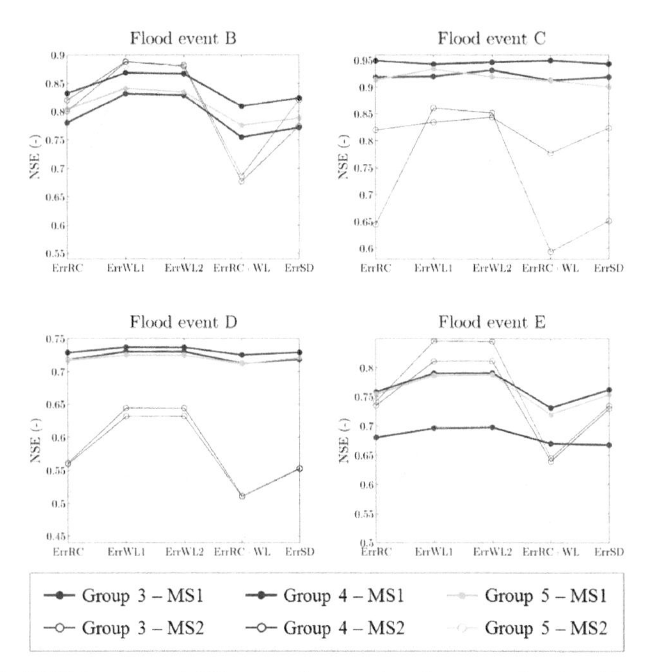

Figure 4.19. NSE in case of different observational errors obtained for flood events B, C, D and E considering sensors located in the group 3, 4 and 5.

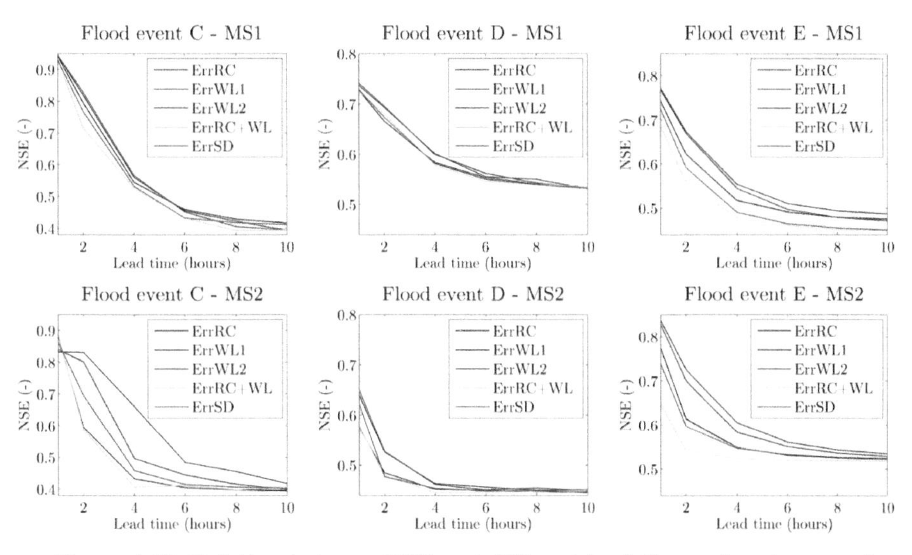

Figure 4.20. Relation between NSE and different lead time values in case of diverse type of observational error during flood events C, D and E considering scenario 6 for both MS1 and MS2.

Moreover, the difference between ErrRC and ErrRC+WL is higher in MS2 than in MS1 for low lead time. It is interesting to observe that NSE values simulated with ErrWL, during flood event C and MS2, are lower than the ones obtained with the other observational errors for 1-hour lead time. However, for higher lead time values, ErrWL1 and ErrWL2 provide the highest model improvements. The difference between NSE obtained using different observational errors tends to be negligible in both MS1 and MS2 in case of high lead time. In particular, for lead time values comparable with the time of concentration of the basin, the NSE values obtained with the different observational errors tend to the one achieved without model update.

4.4.2 Experiment 4.2

In this section, the results obtained in the experiment 4.2 are described considering only SC 8, 9 and 10, since these are the spatial configuration in which the assimilation of distributed static sensors provides the best model improvement. It is worth noting that only flood events A and B are considered in Experiments 4.2, 4.3 and 4.4.

In Figure 4.21, the difference between the outflow hydrograph estimated in the experiment 4.1, considering only fixed ErrRC in time and space for StPh sensors, and experiment 4.2 (from now on called difference-1), is represented. The smaller the value of difference-1, the smaller the sensitivity of the model to assimilation of observations from dynamic sensors. The figure shows that in both the analysed flood events the MS1 is less sensitive to the assimilation of regular observations having variable uncertainty than the MS2.

Physically, this can be due to the attenuation induced by the particular structure of MS1 (sub-catchments connected in series), while in MS2 the high sensitivity of the model is related to the parallel structure of the hydrological models. It can also be seen that the influence of different random situations of StSc sensors is negligible in MS1 but is higher in MS2. In addition, in MS1, the SC10 seems to be more sensitive to the random position of StSc sensors than configuration 9. On the contrary, the value of difference-1 is higher in SC 9 than in SC 10 in case of MS2. A similar situation is found out for the both flood events A and B. In particular, the values of difference-1 in flood event B is smaller than in flood event A due to the different performances of the model without assimilation. In fact, in case of flood event B, additional real-time observations of discharge slightly improved the model results since the model tends to better estimate the observed value of discharge even without assimilation. The difference between assimilation of observation from StPh and StSc in MS1 is negligible considering the flood event B. Opposite situation is however observed for the event A.

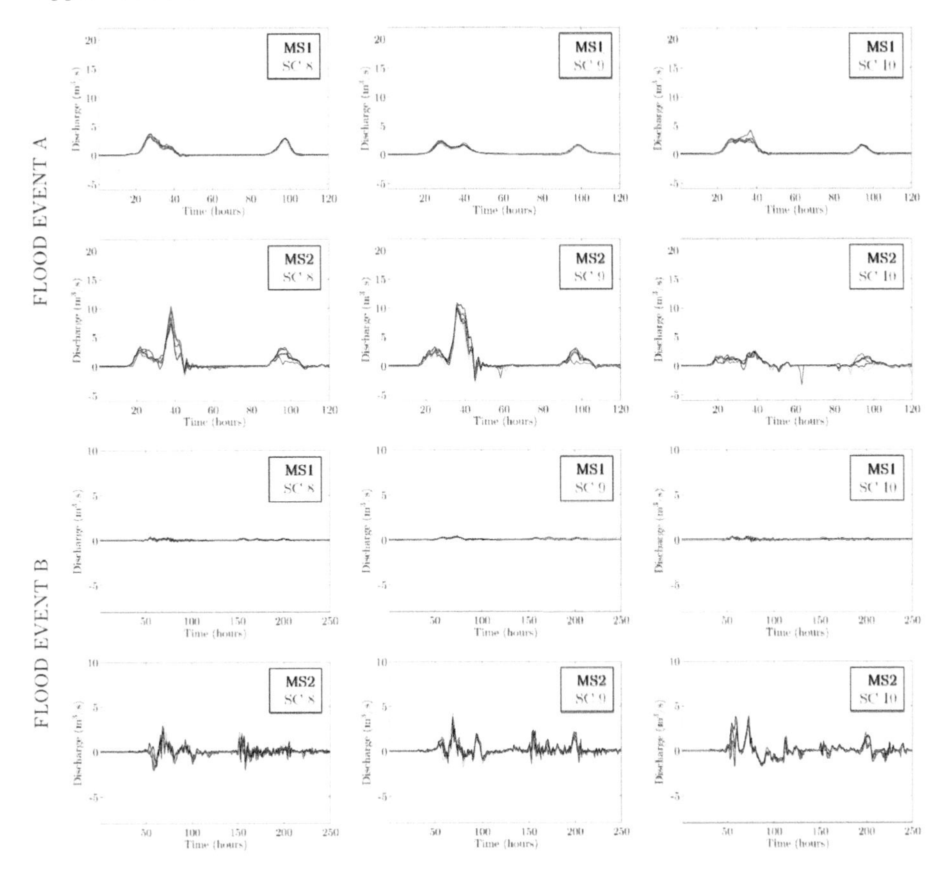

Figure 4.21. Difference between synthetic experiment 4.1 and 4.2 in terms of outflow hydrographs under different configurations of StSc sensors SC (different colour lines).

4.4.3 Experiment 4.3

In experiment 4.3 the StSc sensors with the intermittent observations of discharge within SC 8, 9 and 10 are considered. Figure 4.22 shows the difference between the hydrographs of experiment 4.2 and 4.3 (from now on called difference-2). From Figure 4.22, it can be observed that the values of difference-2 are generally higher than that of difference-1 for the MS1 and MS2.

In particular, for MS1, the model seems to be more sensitive to the intermittent nature of the observations rather than their dynamic behaviour of citizens across

StSc sensors during experiment 4.2. Also in case of intermittent observations, SC 10 is more sensitive than SC 9 considering MS1. Opposite results are achieved with MS2. Similar model performances, with different magnitude of the difference-2, are obtained in case of flood event B.

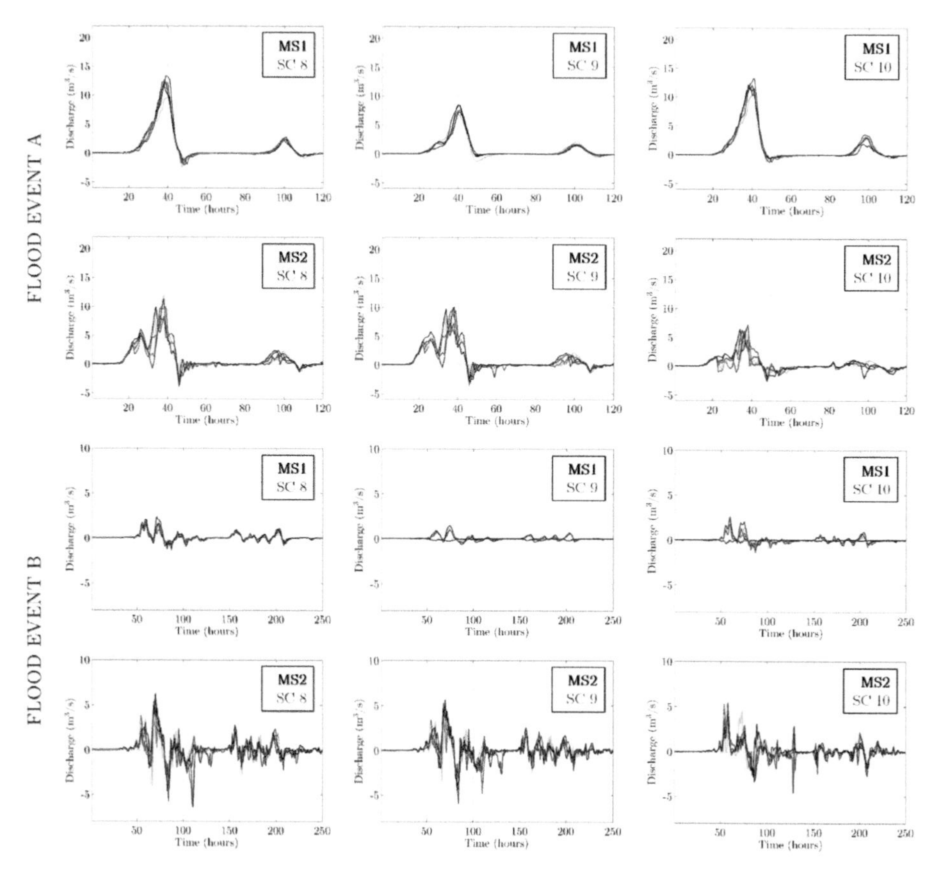

Figure 4.22. Difference between synthetic experiment 4.2 and 4.3 in terms of outflow hydrographs under different configurations of dynamic sensors SC (different colour lines)

In Figure 4.23, the difference between experiment 4.1 and 4.3, called difference-3, are reported to give an additional demonstration of the results previously described. Considering the flood event A, it can be seen how the model outputs are affected by changing from static to intermittent observations of streamflow, mainly in case of SC 10 and SC 9 considering MS1 and MS2 respectively.

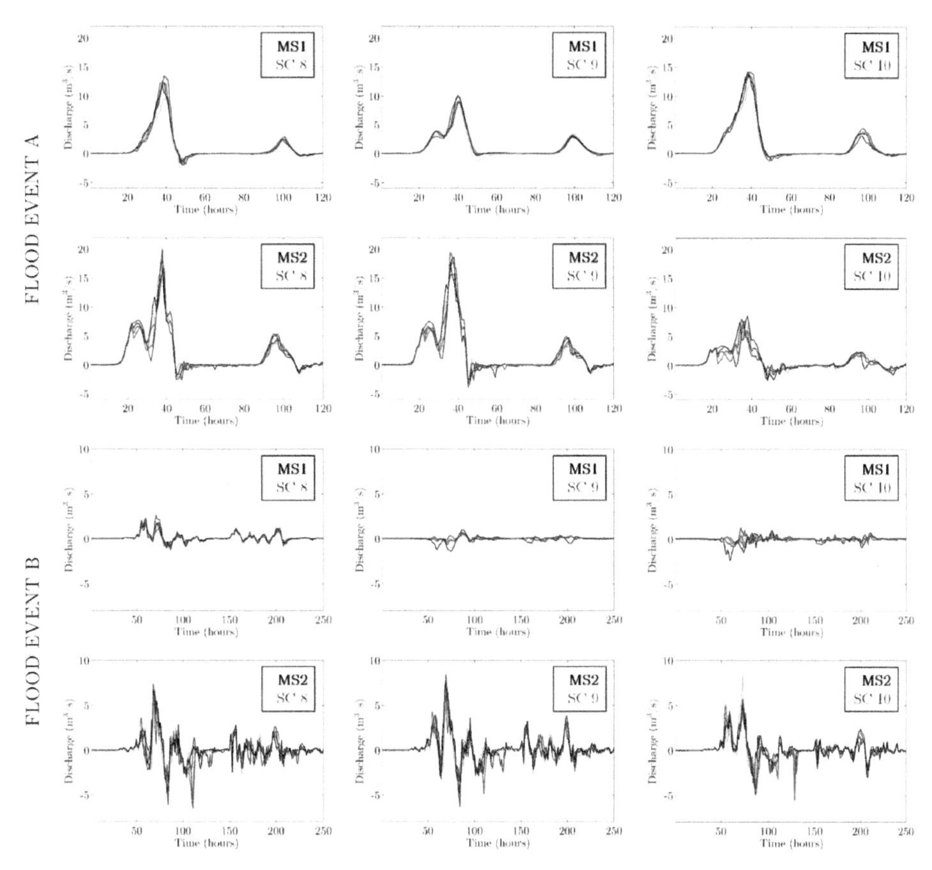

Figure 4.23. Difference between synthetic experiment 4.1 and 4.3 in terms of outflow hydrographs under different configurations of dynamic sensors distributions (different colour lines).

Table 4.4. Nash index values obtained assimilating streamflow observations from SC of sensors in case of two different model structures

	MS1			MS2		
SC	**8**	**9**	**10**	**8**	**9**	**10**
Experiment 4.1	0.77	0.69	0.75	0.82	0.81	0.64
Experiment 4.2	0.74	0.63	0.70	0.56	0.54	0.53
Experiment 4.3	0.58	0.51	0.47	0.45	0.43	0.42

Model performances (NSE values) observed in the three experiments for the three SC are presented in Table 4.4. The results correspond with the ones reported in Figure 4.21, Figure 4.22 and Figure 4.23. For MS1, a small difference in the NSE is seen between experiment 4.1 and experiment 4.2, while experiment 3 shows larger effect of intermittent observations on the model performance. However, in case of MS2, the use of dynamic observations leads to a higher deterioration of model performance, around 30%, of the NSE, than MS1. The uncertainty in simulated discharge is represented by the 90% prediction interval (Figure 4.7 without assimilation and Figure 4.24 with assimilation in case of MS1), i.e. the output that falls into the range between 5% and 95% quantiles of the distribution at each time step of the simulation.

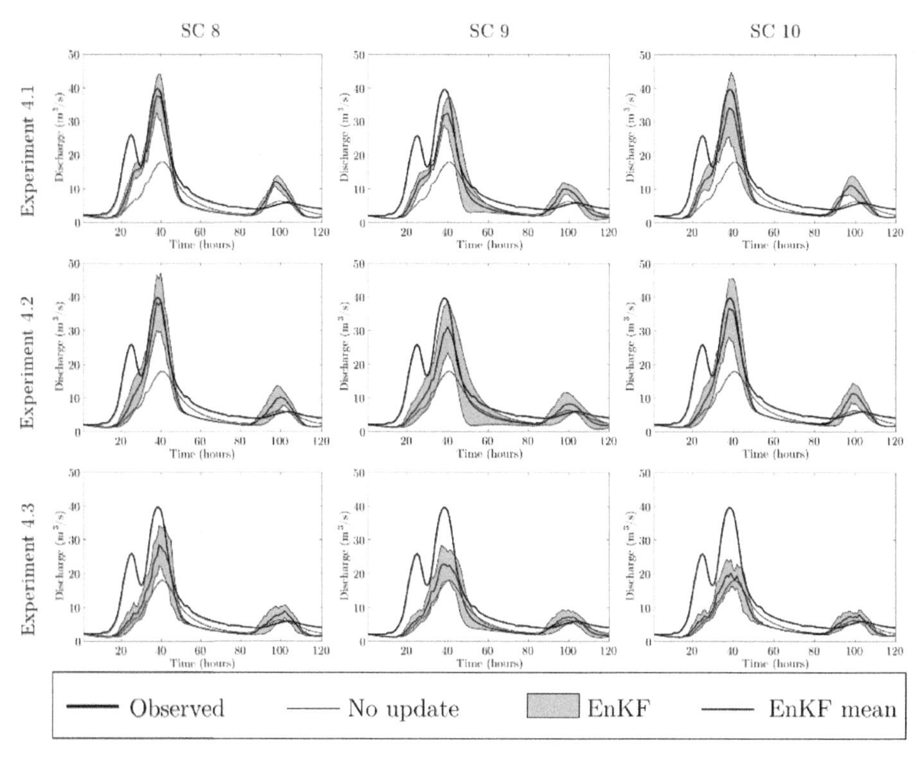

Figure 4.24. Streamflow estimation with 95% prediction interval for the EnKF in case of the SC 8, 9 and 10 for the three proposed experiment considering MS1 during event A

Overall, uncertainty in experiment 4.2 is higher than in experiment 4.1; this can be related to the low accuracy of dynamic sensors which will reflects in a higher uncertainty in the estimation of the peak value of discharge. In case of experiment 4.3, this uncertainty is slightly reduced due to the low value of discharge. The same considerations are valid also for the SC 9 and 10. MS1 tends to provide good average results passing from SC 8 to 10 and then 9. Similar considerations can be drawn also in case of flood event B. In Figure 4.25 the EnKF mean is shown for experiments 4.1, 4.2 and 4.3 considering only SC 9 and SC 10. On the one hand, it can be pointed out that, with StPh and non-intermittent StSc sensors, SC 9 and SC 10 tend to provide the best model improvement in case of MS2 and MS1, respectively. On the other hand, in case of intermittent StSC sensors, SC 9 tends to perform better than SC 10 for MS1.

This can be due to the fact that in SC 9 the number of sensors is higher than in SC 10 (see Table 4.1) and this possibly mitigates the effect of the intermittent nature of streamflow observations.

In addition, this can be explained by the nested structure of MS1. In fact, measurements assimilated in a given catchment on the main river channel, for instance at time t, might improve the output of the downstream sub-catchment at time step t+dt even if in that time step no observations are available at that particular sub-catchment, since it receives updated input from the upstream sub-catchment. Assimilation of discharge information in some sub-catchments can compensate the absence of observations in other sub-catchments due to the semi-distributed nature of the model.

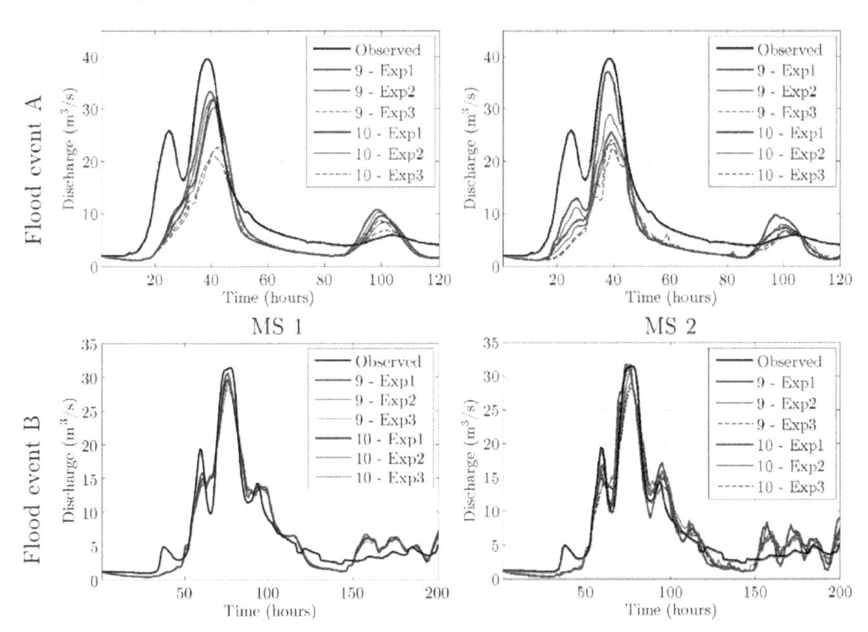

Figure 4.25. Outflow hydrographs in case of implementation of experiments 4.1, 4.2 and 4.33 in SC 9 and 10.

4.4.4 Experiment 4.4

In this experiment a heterogeneous network of sensors, including the StPh sensors and staff gauges (StSc sensors), for several settings of sensors distribution in time and space is considered. The results obtained for the flood event A, in case of MS 1, are reported in Figure 4.26.A. It can be seen that even in case of bad model results due to a possible incorrect input estimation (flood event A), the assimilation of uncertain discharge observations measured at seven staff gauges by StSc sensors could improve the model results (however still with the underestimation of the peak flow for settings 1 and 2). Assimilation of observations coming from the trained volunteers in the peak hours (settings 3 and 4) showed a satisfactory improvement of the discharge hydrograph (higher for the model structure 1 than for the structure 2). Intermittent observations from StSc sensors do not improve the model results in the same way as the observations coming continuously in time from StSc sensors do. Figure 4.26.B confirms that the similar improvements obtained for setting 3 are achieved assimilating observations coming from the optimally located StPh sensors running continuously in time (setting 5). In addition, a combined assimilation of intermittent observations (during daylight hours) with observations from the optimal (setting 7) and non-optimal (setting 12) networks of StPh sensors tends to slightly improve the model output compared to setting 5.

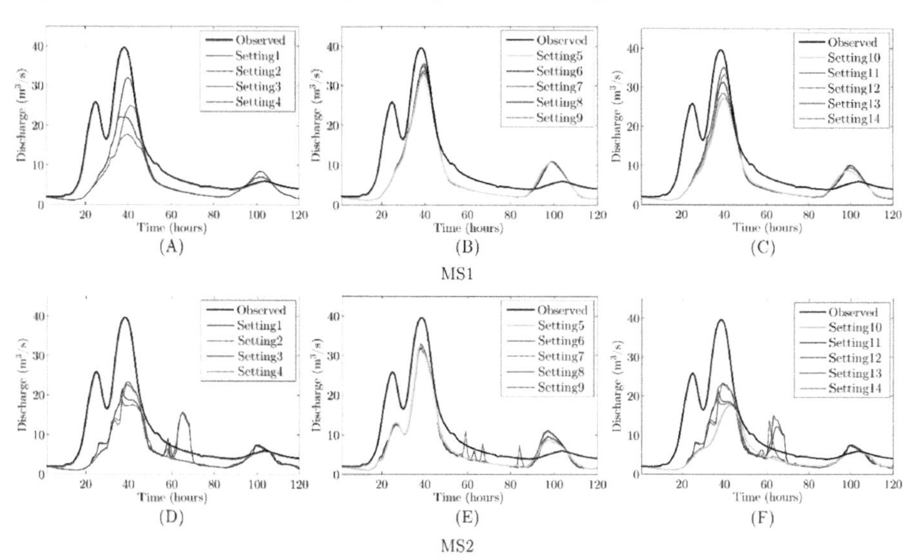

Figure 4.26. Outflow hydrographs resulting from the assimilation of observations from the heterogeneous network of StPh and StSc sensors during the flood event A.

Figure 4.26 demonstrates that, considering this particular type of hydrological model and river basin, the combined assimilation of uncertain streamflow observations, continuous in time, and streamflow observations from a non-optimal network of StPh sensors (setting 13) provides an improvement in the estimation of the outflow hydrograph. In particular, the comparison between settings 1, 3 and 10 shows how an inappropriate distribution of StPh sensors can be replaced by the uncertain streamflow, non-intermittent, observations from StSc. The results also show that a generic non-optimal StPh sensors location (setting 10) can be replaced by a combination of these StPh sensors and intermittent StSc sensors providing streamflow observations in daylight and peak hours (setting 14). It can be also seen, that a combination of StPh and StSc sensors (settings 6, 7, 8 and 9), does not provide any additional improvement to the model if the spatial distribution of these StPh sensors is optimal (setting 5).

Similar conclusions can be draw for MS2, albeit with less improvement in model results. It can be seen how the results obtained from continuous and intermittent StSc sensors (Figure 4.26.D) are very similar to the ones obtained combining non optimal StPh sensors location with the continuous and intermittent StSc sensors (Figure 4.26.F).

Therefore, integrating the observations from continuous and intermittent StSc sensors with the one from optimal StPh sensors (Figure 4.26.E) does not improve the model results for the MS2, in contrast to the MS1 (Figure 4.26.B). These results bring to the conclusion that in order to improve the model performance using continuous and intermittent StSc sensors, not only the locations of such sensors are important but also the model structure plays a crucial role.

4.5 CONCLUSIONS

This chapter evaluated the effect of distributed StPh sensor locations and observation accuracy on the DA performance, and demonstrated to what extent the assimilation of streamflow observations from StSc sensors with varying spatial and temporal coverage could improve flood forecasting. To this end, a methodology was proposed involving updating two simplified semi-distributed hydrological model structures (MS1) nested and (MS2) parallel connection of the sub-catchments) with uncertain discharge observations, distributed in space and intermittent in time. A number of assimilation experiments have been carried out in order to validate the proposed methodology using synthetic discharge series representing the hypothetical uncertain streamflow observations.

From the results of Experiment 4.1 it can be concluded that, for both model structures, assimilation of streamflow observations from static sensors distributed across the whole catchment or on the main river reach provides the best model improvement for the both flood events considered. It is demonstrated that the assimilation of observations coming from the sensors arranged in spatial configuration 2 (streamflow observations at the catchment outlet which is a usual practice in operational DA), does not provide a significant improvement to the discharge estimation. However, flood prediction with MS2 is influenced by the total number of sensors and their locations, while MS1 is only affected by the sensor locations but not by the number of sensors. The DA performances with MS2 are more sensitive to the magnitude of the flood event than MS1. Data assimilation of streamflow observations within two different model structures provided comparable results in terms of forecasting accuracy. In general, for high lead times MS1 is generally better than MS2. It is worth noting that the results we obtained are valid only in the considered case study and for two particular flood events. Considering the variable accuracy of the streamflow measurements, this chapter indicated that considering the error in biased streamflow observations (ErrSD) and error in the water level measurements (ErrWL) has no significant effect on the outflow hydrograph prediction if compared to the results obtained assuming rating curve uncertainty (ErrRC). Both model structures are influenced by different types of observational error only for low lead times while for high lead times NSE tends to be similar in all the assumed observation errors. However, MS2 is more sensitive than MS1 to different types of observational errors for low lead time values.

In case of Experiment 4.2, 4.3 and 4.4, it is found that overall MS1 is less sensitive than MS2 to the assimilation of observations from StSc, while both model structure are influenced by the assimilation of intermittent observations. In particular, it is demonstrated that the influence of different random configurations of StSc sensors is negligible for model structure 1 but is higher for structure 2. This can be related to the way the sub-catchments are hydrologically connected in the two model structures. Finally, the spatial configuration and the temporal schemes of assimilation are developed to imitate a realistic setting when the uncertain citizen-based observations come from reasonable locations at daylight hours. Overall, assimilation of such observations leads to a noticeable improvement of the model performance. Additionally, in case of MS 1, it is shown that the assimilation of measurements during the flow peak (which in the context of citizen observatories can be carried out by trained volunteers) would allow for the further improvements in model accuracy, comparable to the improvements achieved by assimilating the streamflow observations from the optimal network of static sensors. In particular, the results showed that an inappropriate distribution of StPh sensors could be replaced by the uncertain streamflow observations from StSc sensors. For this reason, an non-optimal network of StPh sensors could be integrated with a network of intermittent StSc sensors providing streamflow observations in daylight and peak hours in order to improve model results.

The results of this chapter prove that assimilation of streamflow observations at interior points of the catchment can improve hydrologic models according to the particular location of the static sensors and the hydrologic model structure. In particular, the sensor locations which generate the highest NSE value at the catchment outlet change according to the given flood event. This can be due to the fact different flood events are generated from a different spatial distribution of precipitation. Therefore, designing flow sensor networks considering a longer time series of flood events might be inappropriate since the effect of the single flood event is not considered. For this reason, additional efforts towards the development of techniques to design networks of dynamic low-cost sensors should be carried out. Appropriate definition of the observational accuracy can affect the model performances and the consequent flood forecasting. In addition, it is demonstrated that assimilation of uncertain streamflow observations from StSc sensors can provide similar model improvements to assimilation of streamflow observations coming from a non-optimal network of StPh sensors. This can be a potential application of recent efforts to build citizen observatories of water, which can make the citizens an active part in information capture, evaluation and communication, helping simultaneously to improve modelling-based flood forecasting.

5

ASSIMILATION OF ASYNCHRONOUS DATA IN HYDROLOGICAL MODELS

This chapter describes a methodology to assimilate crowdsourced streamflow observations in hydrological models and shows how this can improve flood prediction. A modified version of the standard Kalman filter approach is implemented and applied to the lumped version of the KMN model for the Brue catchment and the semi-distributed hydrological model implemented in the Bacchiglione catchment, both described in chapter 2. The main disadvantages of crowd-sourced observations are asynchronous arrival frequency and variable accuracy. Realistic (albeit synthetic) streamflow observations are used to represent crowdsourced data, in both case studies. In the Brue catchment the effect of random arrival time and accuracy of crowdsourced observations from a single StSc sensor, is assessed by means of different experimental scenarios. IN addition, the assimilation of asynchronous crowdsourced observations in a heterogeneous network of StPh and StSc sensors is analysed.

This chapter is based on the following peer-reviewed journal publication:

Mazzoleni M., Veerlan M., Alfonso L., Monego M., Norbiato D., Ferri M., and Solomatine D.P. (2015) Can assimilation of crowdsourced streamflow observations in hydrological modelling improve flood prediction?, Hydrology and Earth System Sciences, under review

5.1 INTRODUCTION

Monitoring stations, such as StPh sensors, have been used for decades to properly measure hydrological variables and better predict floods. In recent years, the continued technological improvement has stimulated the spread of low-cost sensors, such as StSc sensors, that allow for measuring hydrological variables in a more distributed and crowdsourced way than the classic StPh sensors allow. The observations from StPh sensors have a well-defined structure in terms of frequency and accuracy. The crowdsourced observations, however, are provided by citizens with varying experience of measuring environmental data and little connections between each other, and the consequence is that the low correlation between the measurements might be observed. For this reason, these observations can be defined as asynchronous because do not have predefined rules about the arrival frequency (the observation might be sent just once, occasionally or at irregular time steps which can be smaller than the model time step).

In operational hydrology practice so far, the added value of asynchronous crowdsourced information it is not integrated into the forecasting models but just used to compare the model results with the observations in a post-event analysis. One reason can be related to the intrinsic variable accuracy, due to the lack of confidence in the data from such heterogeneous sensors, and the variable life-span of the observations. In the previous chapter the effects of distributed synthetic streamflow observations having synchronous intermittent temporal behaviour and variable accuracy in a semi-distributed hydrological model is analysed.

A possible solution to handle asynchronous observations in time with EnKF is to assimilate them at the moments coinciding with the model time steps (Sakov et al., 2010). However, as these authors mention, this approach requires the disruption of the ensemble integration, the ensemble update and a restart, which may not feasible for large-scale forecasting applications.

Continuous approaches, such as 3D-Var or 4D-Var methods, are usually implemented in oceanographic modelling in order to integrate asynchronous observations at their corresponding arrival moments (Derber and Rosati, 1989; Macpherson, 1991; Huang et al., 2002; Ragnoli et al., 2012). In fact, oceanographic observations are commonly collected at not pre-determined, or asynchronous, times. For this reason, in variational data assimilation, the past asynchronous observations are simultaneously used to minimise the cost function that measures the weighted difference between background states and observations over the time interval, and identify the best estimate of the initial state condition (Ide et al., 1997; Li and Navon, 2001; Drecourt, 2004). In addition to the 3D-Var and 4D-Var methods,

Hunt et al. (2004) proposed a Four Dimensional Ensemble Kalman Filter (4DEnKF) which adapts EnKF to handle observations that have occurred at non-assimilation times. In this method the linear combinations of the ensemble trajectories are used to quantify how well a model state at the assimilation time fits the observations at the appropriate time. Furthermore, in case of linear dynamics 4DEnKF is equivalent to instantaneous assimilation of the measured data (Hunt et al., 2004). Similarly to 4DEnKF, Sakov et al. (2010) proposed the Asynchronous Ensemble Kalman Filter (AEnKF), a modification of the EnKF, mainly equivalent to 4DEnKF, used to assimilate asynchronous observations (Rakovec et al., 2015). Contrary to the EnKF, in the AEnKF current and past observations are simultaneously assimilated at a single analysis step without the use of adjoint model. Yet another approach to assimilate asynchronous observations in models is the so-called First-Guess at the Appropriate Time (FGAT) method (Massart et al., 2010). Like in 4D-Var, the FGAT compares the observations with the model at the observation time. However, in FGAT the innovations are assumed constant in time and remain the same within the assimilation window (Massart et al., 2010).

Having reviewed all the described approaches, a straightforward and pragmatic method, similar to the AEnKF, is used to assimilate the asynchronous crowdsourced observations. This is due to the linearity of the hydrological models implemented.

The main objective of this chapter is to assess the potential use of crowdsourced observations within hydrological modelling. In particular, the specific objectives are to a) assess the influence of different arrival frequency of the crowdsourced observations and their related accuracy on the assimilation performances in case of a single StSc sensor; b) to integrate the distributed low-cost StSc sensors with a single StPh sensor to assess the improvement in the flood prediction performances in an EWS. It is worth noting that in Mazzoleni et al. (2015b) additional nine flood events and two case studies are considered in order to further validate the results achieved in this Thesis.

5.2 METHODOLOGY

In order to assimilate asynchronous CS flow observations within the lumped hydrological model implemented for in the Brue catchment (see section 2.2.2 on page 30) and the semi-distributed model used in Bacchiglione (see section 2.3.2 on page 37), of a modification of the standard KF is used. Also in this chapter, as in case of Chapter 4, synthetic realistic streamflow observations, asynchronous in time and with random accuracy, are used due to the fact that CS observations are not available at the time of this study. In particular, in the Bacchiglione catchment, the StSc sensors are being recently installed in the summer of 2014 within the framework of the WeSenseIt project. Below, the description of the DA method used to assimilate asynchronous observations, definition of the observation and model error and the estimation of realistic synthetic observations are reported.

5.2.1 Assimilation of asynchronous observations

In most of the hydrological applications of DA, observations from StPh sensors are integrated into water models at a regular, synchronous, time step. However, as showed in Figure 1.7, a StSc sensor can be used by different operators, having different accuracy, to measure water level at a specific point. For this reason, StSc sensors provide CS observations which are asynchronous in time and with a higher degree of uncertainty than the one of observations from StPh sensors. In particular, CS observations have three main characteristics: a) irregular arrival frequency (asynchronicity); b) random accuracy; c) random number of observations received by the static device within two model time steps.

As described in Introduction of this chapter, various methods have been proposed in order to include asynchronous observations in models. Having reviewed them, in this thesis a somewhat simpler DA approach for integrating Crowdsourced Observations into hydrological models (DACO) is proposed. This method is based on the KF approach described in Chapter 2 and on the assumption that the change in the model states and in the error covariance matrices within the two consecutive model time steps t_0 and t (observation window) is linear, while the inputs are assumed constant. All the observations received during the observation window are assimilated in order to update the model states and output at time t. Therefore, assuming that one observation would be available at time t_0^*, the first step of such a filter (box A in Figure 5.1) is the definition of the model states and error covariance matrix at t_0^* as:

$$\mathbf{x}_{t_0^*}^- = \mathbf{x}_{t_0}^+ + \left(\mathbf{x}_t^- - \mathbf{x}_{t_0}^+\right) \cdot \frac{t_0^* - t_0}{t - t_0} \tag{5.1}$$

$$\mathbf{P}_{t_0^*}^- = \mathbf{P}_{t_0}^+ + \left(\mathbf{P}_t^- - \mathbf{P}_{t_0}^+\right) \cdot \frac{t_0^* - t_0}{t - t_0} \tag{5.2}$$

The second step (B in Figure 5.1) is the estimation of the updated model states and error covariance matrix, as the response to the streamflow observation $Q_{t_0^*}^o$. The estimation of the posterior values of $\mathbf{x}_{t_0^*}^-$ and $\mathbf{P}_{t_0^*}^-$ is performed by Eqs. (3.9) and (3.10) respectively. The Kalman gain, estimated by Eq. (3.8), where the prior values of model states and error covariance matrix at t_0^* are used. Knowing the posterior value of $\mathbf{x}^+_{t_0^*}$ and $\mathbf{P}^+_{t_0^*}$ it is possible to predict the value of states and covariance matrix at one model step ahead, t^* (C in Figure 5.1) using the model forecast equations Eqs. (3.6) and (3.7).

The last step (D in Figure 5.1) is the estimation of the interpolated value of $\mathbf{x}$ and $\mathbf{P}$ at time step t. This is performed by means of a linear interpolation between the current values of $\mathbf{x}$ and $\mathbf{P}$ at t_0^* and t^*:

$$\tilde{\mathbf{x}}_t^- = \mathbf{x}_{t_0^*}^- + \left(\mathbf{x}_{t^*}^- - \mathbf{x}_{t_0^*}^+\right) \cdot \frac{t - t_0^*}{t^* - t_0^*} \tag{5.3}$$

$$\tilde{\mathbf{P}}_t^- = \mathbf{P}_{t_0^*}^- + \left(\mathbf{P}_{t^*}^- - \mathbf{P}_{t_0^*}^+\right) \cdot \frac{t - t_0^*}{t^* - t_0^*} \tag{5.4}$$

The symbol ~ is added on the new matrices $\mathbf{x}$ and $\mathbf{P}$ in order to differentiate them from the original forecasted values in t. Assuming that the new streamflow observation is available at an intermediate time t_1^* (between t_0^* and t), the procedure is repeated considering the values at t_0^* and t as for the linear interpolation. Then, in case when no more observations are available, the updated value of $\mathbf{x}^+(t)$ is used to predict the model states and output at $t+1$ (Eqs.(3.6) and (3.7)).

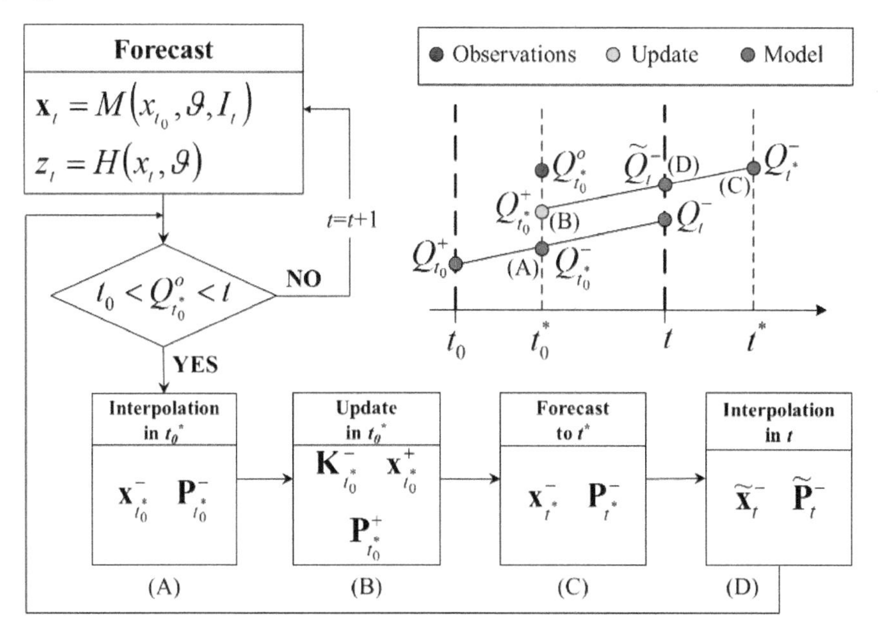

Figure 5.1 Graphical representation of the DACO method proposed in this chapter to assimilate CS asynchronous observations

Finally, in order to account for the intermittent behaviour of such observations, the approach proposed in Chapter 4 is applied. In this method, the model states matrix **x** is updated and forecasted when observations are available, while without observations the model is run using Eq. (3.6) and covariance matrix **P** propagated at the next time step using Eq.(3.7).

5.2.2 Observation and model error

In this section, the uncertainty related to the streamflow observations and model error are characterised. The observational error is assumed to be the normally distributed noise with zero mean and given standard deviation as described in Eq.(4.4). Due to the unpredictable accuracy in the CS the coefficient is assumed to be a random stochastic variable between 0.1 and 0.3 (Mazzoleni et al., 2015a, 2015b). In this Chapter, is considered changing only in time since the location of the StSc is a priori assigned, in contrast to Chapter 5. Cortes et al. 2014 argued (and this is a reasonable suggestion) that the uncertainty of a measurement provided by a well-trained technician is smaller than the one coming from a normal citizen. For this reason it is assumed that the maximum value of is three times higher than the uncertainty coming from the physical sensors (Mazzoleni et al., 2015a, 2015b).

In case of the lumped model implemented in the Brue catchment, the covariance matrix **S** is considered as a stationary diagonal matrix having fixed value equal to 1. With such a small value, the model is assumed to be more accurate than the observations, in case of high flow. In this way, it is possible to assess the additional value provided by the assimilation of CS observations. Usually, the model error is estimated as the diagonal matrix having constant value equal to the standard deviation between simulated and observed streamflow at the station of Vicenza (see hydrographs in Figure 2.12). However, in this chapter, in order to evaluate the effect of assimilating CS observations, the model is considered more accurate than the observations and, a covariance matrix **S** with diagonal values of 10^2 is considered.

5.2.3 Generation of synthetic observations

Realistic streamflow observations are generated for the Brue catchment considering the observed streamflow values at the outlet section of the catchment during two flood events (see Figure 2.4) occurred from 28/10/1994 to 16/11/1994 (flood event 1) and from 14/01/1995 to 08/08/1995 (flood event 2). For this reason, observed hourly streamflow observations at the catchment outlet are interpolated to represent observations coming at arrival frequency higher than hourly. Different experiment are run and described more in detailed in the next sections.

For the Bacchiglione catchment, the realistic CS observations are based on model results. In fact, streamflow values obtained using measured precipitation in May 2013 flood event as input in the hydrological model (post-event simulation) are used as hourly streamflow observations. Then, like in case of the Brue catchment, interpolated streamflow values, having different accuracy and arrival time, are used to represent CS observations at the StSc sensors located at the outlet of the sub-catchments A, B and C of the Bacchiglione catchment (see Figure 2.7). Forecasted precipitation value are used as input in the hydrological model to estimate the simulated streamflow.

5.3 EXPERIMENTAL SETUP

In this section, two sets of experiments are performed in order to test the proposed method and assess the benefit to integrate CS observations, asynchronous in time and with variable accuracy, in real-time flood forecasting.

In the first set of experiments, called "Experiments 5.1", assimilation of streamflow observations at one StSc sensor location is carried out to understand the sensitivity of the employed hydrological model (KMN) under various scenarios of such observations.

In the second set of experiments, called "Experiments 5.2", the distributed observations coming from StPh and StSc sensors, at four locations within the Bacchiglione catchment, are considered, with the aim of assessing the improvement in the flood forecasting accuracy.

5.3.1 Experiment 5.1: Observations from a single static social (StSc) sensor

The focus of Experiment 1 is to study the performance of the hydrological model (KMN) assimilating CS observations, having lower arrival frequencies than the model time step and random accuracies, coming from a social sensor located in a specific point of the Brue catchment. To analyse all possible combinations of arrival frequency, number of observations within the observation window (1 hour) and accuracy, a set of scenarios are considered (Figure 5.2), changing from regular arrival frequency of observations with high accuracy (scenario 1) to random and chaotic asynchronous observations with variable accuracy (scenario 11).

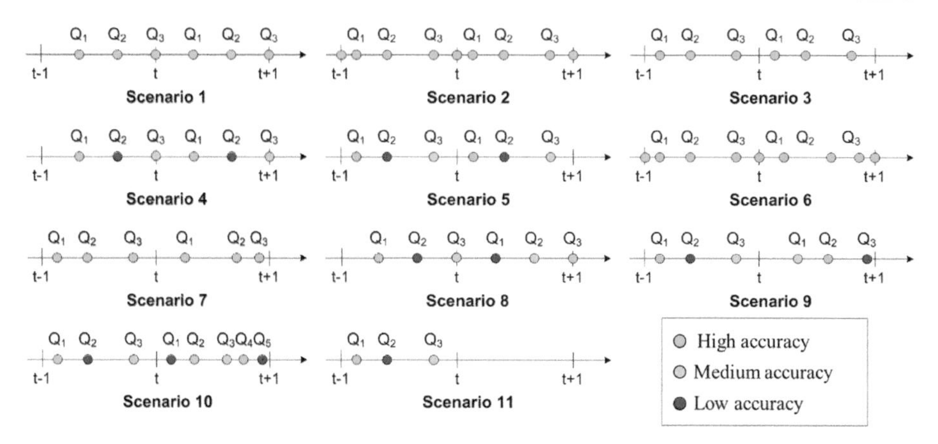

Figure 5.2. The experimental scenarios representing different configurations of arrival frequency, number and accuracy of the streamflow observations

Each scenario is repeated varying the number of observations from 1 to 100. It is worth noting that in case of one observation per hour and regular arrival time, scenario 1 corresponds to the case of StPh sensors with an observation arrival frequency of one hour. Scenario 2 corresponds to the case of observations having fixed accuracy (α equal to 0.1) and irregular arrival moments, but in which at least one observation will coincide with the model time step. In particular, scenario 1 and 2 are exactly the same in case of one observation available within the observation window since it is assumed that the arrival frequency of that observation has to coincide with the model time step. On the other hand, the arrival frequency of the observations in scenario 3 is assumed to be random, and observations might not arrive at the model time step.

Scenario 4 considers observations with regular frequency and random accuracy at different moments within the observation window, whereas in scenario 5 observations have irregular arrival frequency and random accuracy. In all the previous scenarios the arrival frequency, the number and accuracy of the observations are assumed to be periodic, i.e. repeated between consecutive observation windows along all the time series. However such periodic repetitiveness might not occur in real-life, and for this reason, a non-periodic behaviour is assumed in scenarios 6, 7, 8 and 9. The non-periodicity assumptions of the arrival frequency and accuracy are the only factors that differentiate scenarios 6, 7, 8 and 9 from the scenarios 2, 3, 4, and 5 respectively. In addition, the non-periodicity of the number of observations within the observation window is introduced in scenario 10.

Finally, in scenario 11 the observations, in addition to all the previous characteristics, might have an intermittent behaviour, i.e. not being available for one or more observation windows.

5.3.2 Experiments 5.2: Observations from distributed static physical (StPh) and static social (StSc) sensors

The main goal of Experiments 5.2 is to understand the contribution of CS observations to the improvement of the flood prediction at a specific point of the catchment, in this case at PA. The synthetic observations using in this experiment have the same characteristics reported in scenarios 10 and 11, in Experiments 5.1. Streamflow observations from StPh sensors are assumed to be synchronous and assimilated in the hydrological model of AMICO system at an hourly frequency, while CS observations from StSc sensors are considered asynchronous (higher and irregular frequency) and are assimilated using the DACO method previously described. The updated hydrograph estimated by the hydrological model is used as the input into Muskingum-Cunge model used to propagate the flow downstream, to the gauged station at PA, Vicenza. Five different experimental settings are introduced, and represented in Figure 5.3, corresponding to different types of sensors used.

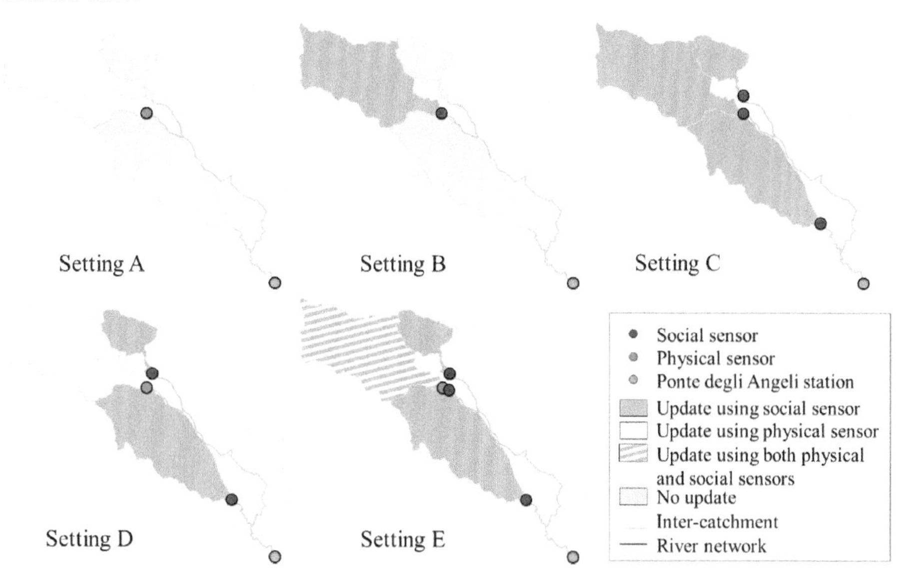

Figure 5.3. Different experimental settings implemented within the Bacchiglione catchment during Experiment 5.2

Firstly, only the observations coming from the StPh sensor at the Leogra sub-catchment are used to update the hydrological model of catchment B (setting A). Secondly, in setting B the model improvement in case of assimilation of CS observations at the same location of setting A is analysed. In setting C only the distributed CS observations within the catchment are assimilated into the hydrological model. Then, setting D accounts for the integration of CS and physical observations, contrary to the setting C where the StPh sensors is dropped in favour of the StSc sensor at Leogra. Finally, setting E consider the complete integration between StPh and StSc sensors in Leogra, Timonchio and Orolo sub-catchments.

5.4 Results and discussion

5.4.1 Experiment 5.1

The observed and simulated hydrographs at the outlet section of the Brue catchment with and without the model update (considering hourly streamflow observations) are reported in Figure 5.4 for two different flood events. As expected, it can be seen that the updated model tends to better represent the flood events than model without updating.

The results of scenario 1, having from 1 to 30 observations within the observation window, are represented in Figure 5.5. As it can be seen, increasing the number of CS observations within the observation window results in the improvement of the NSE for different lead time values, but it becomes negligible, in average, for more than five observations.

This means that the additional CS observations do not add information useful for improving the model performance. This asymptotic behavior when extra information is added has also been observed using other metrics by Krstanovic and Singh (1992), Ridolfi et al. (2014), Alfonso et al. (2013), among others. In all flood events, similar trends of the NSE are found. However, it is not possible to define a priori number of observations needed to improve model. In fact, after a threshold number of observations (five for flood event 1 and fifteen for flood event 2), NSE asymptotically approaches to a certain value meaning that no improvement is achieved with additional observations. However, the only difference between the flood events is that such asymptotic NSE values are different because model performances can change according to the considered flood events. In case of this case study and during these five flood events, it can be seen that an indicative value of 10 observations can be considered in average to achieve a good model improvement.

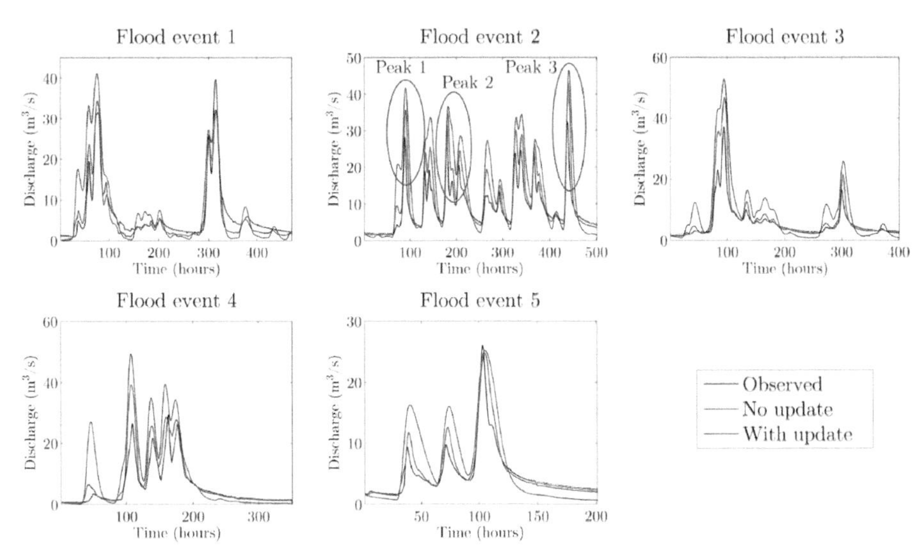

Figure 5.4. The observed and simulated hydrographs, with and without assimilation, for five considered flood events in the Brue catchment. In case of event 2, three flood peaks are considered and analysed below

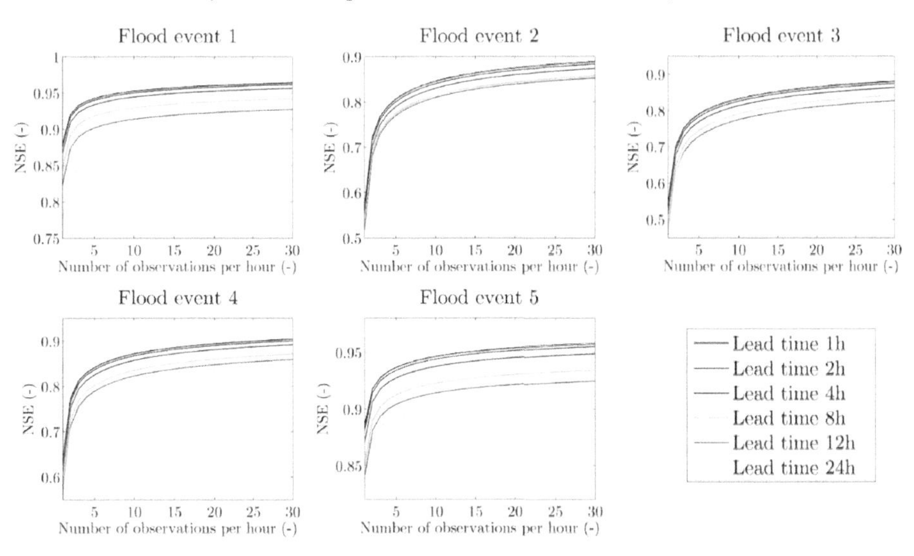

Figure 5.5. Model improvement in case of different lead times assimilating streamflow observations according to scenario 1

The same type of analysis is performed with the scenarios 2 to 9 (Figure 5.6 in case of flood event 1). The results obtained in Figure 5.6 show that in case of irregular arrival frequency (scenarios 2 and 3) the NSE is higher than in scenarios 4 and 5, where observations vary in accuracy. These results point out that the model performance is more sensitive to the accuracy of the observations than to the moment in time at which the streamflow observations become available. However, it can be observed that from scenarios 2 to 5 that the trend it is not as smooth as the one obtained with scenario 1. This can be related to the fact that NSE may vary with varying arrival frequency and observations accuracy. In fact, in scenario 1 the arrival frequency is set as regular for different model runs, so the moments in which the observations became available will always be the same for any model run.

On the other hand, in the other scenarios, the irregular moment in which the observation becomes available within the observation window is randomly selected and is changing according to the different model runs. This means that for a given number of observations (for example 5), the five observations will arrive at different moments, for different model runs, and this results in five different values of NSE. A smooth trend is also obtained for scenarios 6, 7, 8 and 9 but this is related to the periodic behaviour of the observations as explained below.

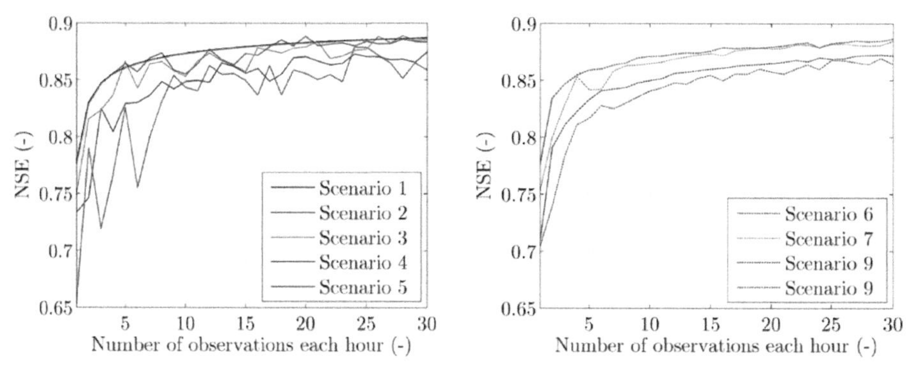

Figure 5.6. Model improvement during flood event 1 (24h lead time), assimilating diverse values of streamflow observations according to the experimental scenarios from 1 to 9 with observations with (a) periodic behaviour and (b) non-periodic behaviour

In order to remove the random behaviour related to the irregular arrival frequency and observation accuracy, different model runs (100 in this case) are carried out, assuming different random values of arrival and accuracy (coefficient α) during each model run, for a given number of observations and lead time. The NSE value is estimated for each model run, so μ(NSE) and σ(NSE) represent the mean and standard deviation of the different values of NSE.

From Figure 5.7 it can be seen that, overall, assimilation of crowdsourced observations improves model performances in all the considered flood events. In case of scenarios 2 and 3 (represented using warm, red and orange, colors in Figure

5.7 and Figure 5.8 for lead time equal to 24 hours), i.e. random arrival frequency with fixed/controlled accuracy, the average values of NSE, μ(NSE), are smaller but comparable with the ones obtained in case of scenario 1 for all the considered flood events. In particular, scenario 3 has lower μ(NSE) than scenario 2. This can related to the fact that both scenarios has random arrival frequency, however, in scenario 3 observations are not provided at the model time step, as opposed to scenario 2. From Figure 5.8, higher values of σ(NSE), the standard deviation of the NSE sample, can be observed in case of scenario 3. Scenario 2 has the lowest standard deviation for low values of discharge observations due to the fact that the arrival frequency has to coincide with the model time step and this tends to stabilize the NSE. It is worth nothing that scenario 1 has null standard deviation due to the fact that observations are assumed coming at the same moment with the same accuracy for the all 100 model runs. In scenario 4, represented using cold blue color, observations are considered coming at regular time steps but having random accuracy. Figure 5.7 shows that μ(NSE) values are lower in case of scenario 4 rather than scenarios 2 and 3. This can be related to the higher influence of observations accuracy if compared to arrival frequency. Such results can be observed in Figure 5.8 as well. In fact, variability in the model performances is higher, especially for low values of CS observations, in scenario 4 than scenarios 2 and 3. The combined effects of random arrival frequency and observation accuracy is represented in scenario 5 using a magenta color (i.e. the combination of warm and cold colors) in Figure 5.7 and Figure 5.8. As expected, this scenario is the one with the lower and higher values of μ(NSE) and σ(NSE), respectively, if compared to the previous ones.

The remaining scenarios, from 6 to 9, are equivalent to the ones from 2 to 5 with the only difference that are non-periodic in time. For this reason, in Figure 5.7 and Figure 5.8, scenarios from 6 to 9 have the same color of scenarios 2 to 5 but indicated with dashed line in order to underline their non-periodic behavior. Overall it can be observed that non-periodic scenarios have similar μ(NSE) values to their corresponding periodic scenario. However, their smoother μ(NSE) trends are due to lower σ(NSE) values which means that model performances are less dependent to the non-periodic nature of the CS observations than their period behavior. Overall, σ(NSE), tend to decrease for the high number of observations. Table 5.1 shows the NSE values and model improvement obtained for the different experimental scenarios during the five flood events. It can be seen that lower improvement is achieved in case of scenarios where arrival frequency is random and accuracy fixed if compared to those scenarios (4, 5, 8 and 9) where arrival frequency is regular and accuracy is random. In addition, flood events with high NSE values even without update tends to achieve the asymptotic values of NSE for small number of observations (e.g. flood event 1 and 5), while more observations are needed for flood events having low NSE without update (flood event 2, 3 and 4).

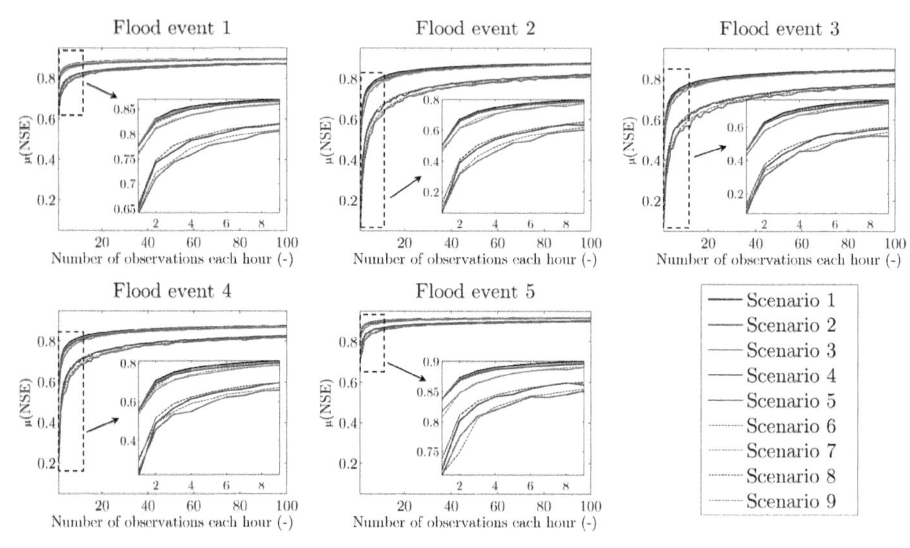

Figure 5.7. Dependency of μ(NSE) and σNSE) on the number of observations, for the scenarios 2, 3, 4, 5, 6, 7, 8 and 9 in case of flood event 1

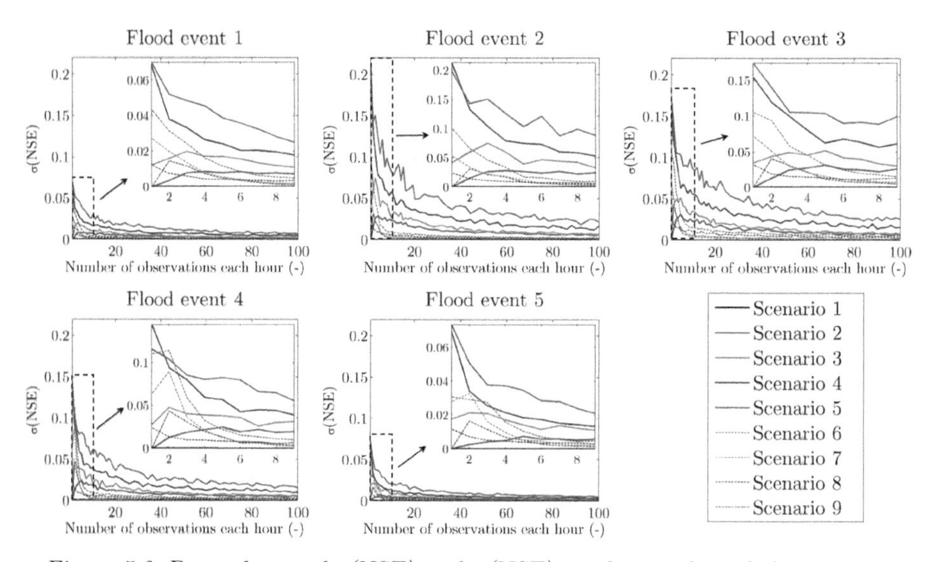

Figure 5.8. Dependency of μ(NSE) and σ(NSE) on the number of observations, for the scenarios 2, 3, 4, 5, 6, 7, 8 and 9 in case of flood event 2

Table 5.1 NSE values in case of different experimental scenarios during the 5 events

	Scenario	1	2	3	4	5	6	7	8	9
Event 1	1obs	0.778	0.778	0.763	0.652	0.643	0.778	0.760	0.659	0.646
	100obs	0.896	0.895	0.895	0.874	0.871	0.895	0.895	0.875	0.872
	Imp (%)	**0.131**	**0.130**	**0.147**	**0.254**	**0.262**	**0.130**	**0.150**	**0.247**	**0.259**
Event 2	1obs	0.504	0.504	0.475	0.063	0.075	0.504	0.459	0.126	0.101
	100obs	0.876	0.872	0.871	0.820	0.816	0.874	0.873	0.821	0.814
	Imp (%)	**0.426**	**0.423**	**0.455**	**0.923**	**0.908**	**0.424**	**0.474**	**0.847**	**0.876**
Event 3	1obs	0.463	0.463	0.434	0.082	0.115	0.463	0.435	0.142	0.129
	100obs	0.847	0.844	0.842	0.778	0.764	0.844	0.842	0.778	0.771
	Imp (%)	**0.454**	**0.452**	**0.484**	**0.895**	**0.850**	**0.452**	**0.483**	**0.818**	**0.832**
Event 4	1obs	0.561	0.561	0.551	0.222	0.246	0.561	0.536	0.293	0.308
	100obs	0.874	0.871	0.871	0.829	0.821	0.872	0.871	0.829	0.823
	Imp (%)	**0.358**	**0.356**	**0.368**	**0.732**	**0.701**	**0.356**	**0.384**	**0.646**	**0.626**
Event 5	1obs	0.838	0.838	0.818	0.720	0.711	0.838	0.808	0.737	0.713
	100obs2	0.918	0.917	0.916	0.904	0.901	0.917	0.916	0.904	0.901
	Imp (%)	**0.087**	**0.086**	**0.107**	**0.203**	**0.211**	**0.086**	**0.119**	**0.184**	**0.209**

The combination of all the previous scenarios is represented by scenario 10 considering the number of CS observations changing at in each observation windows. In scenario 11 the intermittent nature of CS observations is accounted as well. The μ(NSE) and σ(NSE) values of these scenarios obtained for the five considered flood events are showed in Figure 5.9. It can be observed that scenarios 10 tends to provide higher μ(NSE) and lower σ(NSE) values, for a given flood event, if compared to scenarios 11. However, the assimilation of irregular number of observations in scenario 10 in each observation window seems to provide the same μ(NSE) than the ones obtained with scenario 9. One the main outcome is that the intermittent nature of the observations (scenario 11) induces a drastic reduction of the NSE and an increase in its noise in both considered flood events.

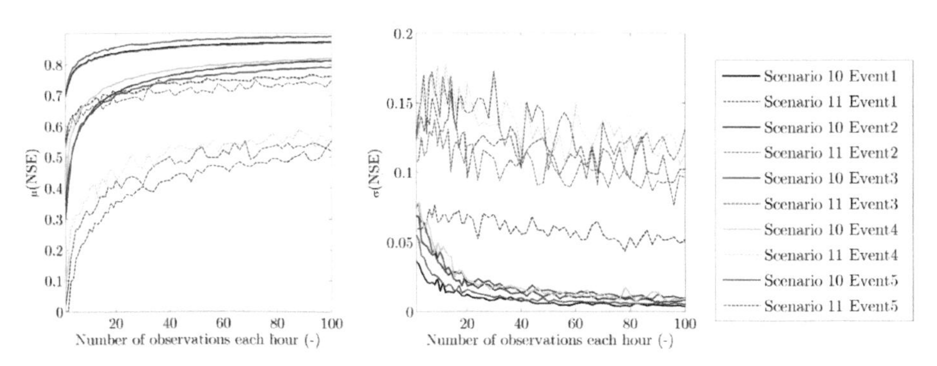

Figure 5.9. Dependency of the (NSE) and (NSE) on the number of observations, for the scenarios 10 and 11 in case of flood events 1 (a) and 2 (b)

In the previous analysis, model improvements are expressed only in terms of NSE. However, statistics such as NSE only explain the overall model accuracy and not to real increases/decreases in prediction error. Therefore, increases in model accuracy due to the assimilation of crowdsource observations have to be presented in different ways as increased accuracy of flood peak magnitudes and timing. For this reasons, additional analyses are carried out to assess the change in flood peak prediction considering 3 peaks occurred during flood event 2 (see Figure 5.4). Error in the flood peak timing and intensity is estimated using Err_t and Err_I equal to:

$$Err_t = t_p^o - t_p^s \tag{5.5}$$

$$Err_I = \frac{Q_p^o - Q_p^s}{Q_p^o} \tag{5.6}$$

Where t_p^o and t_p^s are the observed and simulated peak time (hours), while Q_p^o and Q_p^s are the observed and simulated peak intensity (m^3/s). From the results in Figure 5.9 considering 12-hours lead time, it can be observed that, overall, errors reduction in peak prediction is achieved for increasing number of CS observations. In particular, assimilation of CS observations has more influence in the reduction of the peak intensity rather than peak timing. In fact, a small reduction of Err_t of about 1 hour is obtained even increasing the number of observations. In both Err_I and Err_t the higher error reduction is obtained considering fixed observation accuracy and random arrival frequency (e.g. scenarios 1, 2, 3, 6 and 7). In fact, smaller Err_I error values are obtained in case of scenario 1, while scenarios 5 and 9 is the one that shows the lowest improvement in terms of peak prediction. These conclusions are very similar to the previous ones obtained analyzing only NSE as model performance measures.

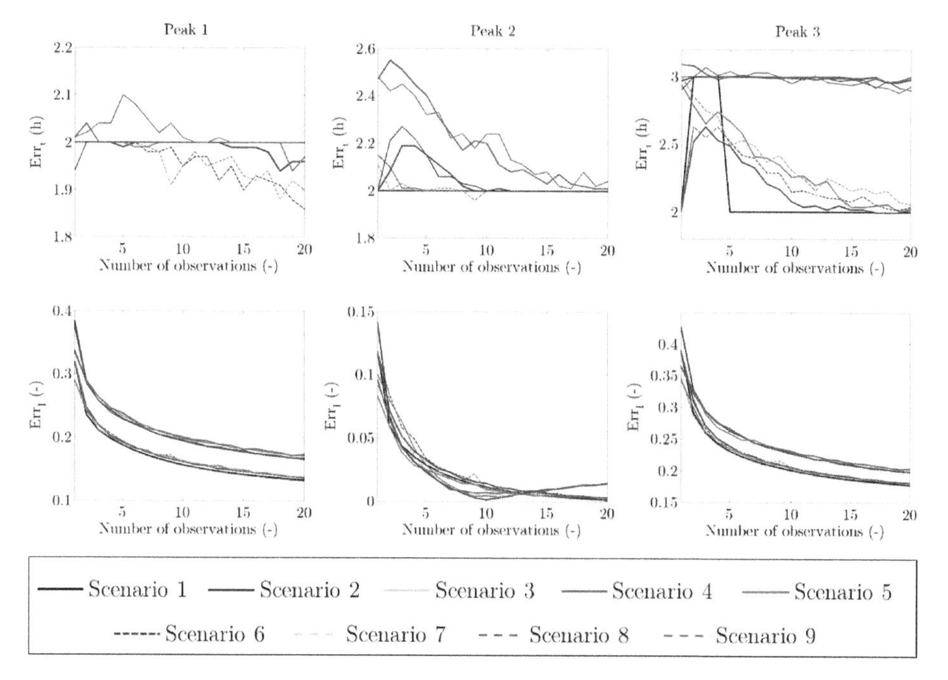

Figure 5.10. Representation of Err_t and Err_I as function of the number of CS observations and experimental scenarios for 3 different peaks occurred during flood event 2

These conclusions are very similar to the ones obtained analysing only NSE as model performance measures. This can be related to the linear nature of the model and the consequent DA approach used in this work.

5.4.2 Experiment 5.2

The physical and CS observations are assimilated in order to improve the poor model prediction in Vicenza affected by the underestimated estimation of the 3-days rainfall forecast used as normal input in flood forecasting practice in this area. Three different flood events occurred in the Bacchiglione catchment are used within Experiments 2. Scenarios 10 and 11, described in the previous sections, are used in this experiment in order to represent an irregular and random behaviour of the CS observations.

The results of this analysis are shown in Figure 5.11. Different model runs (100) are performed for the Leogra sub-catchment, Figure 5.11, to account for the effect induced by the random arrival frequency and accuracy of the CS observations within the observation window as described above. It can be seen that the assimilation of observations from StPh sensors provides a better flood prediction at the Leogra catchment if compared to the assimilation of a small number of CS observations from StSc sensors. In particular, Figure 5.11a and b show that the same NSE values achieved with assimilation of physical observations (hourly frequency and high accuracy) can be obtained by assimilating between 10 and 20 CS observations per hour. However, the overall reduction of NSE in case of intermittent observations is such that even with a high number of observations (even higher than 50 per hour) the NSE is always lower than the one obtained assimilating physical observations for any lead time. Figure 5.11c and d show analogous results expressed in terms of different lead times.

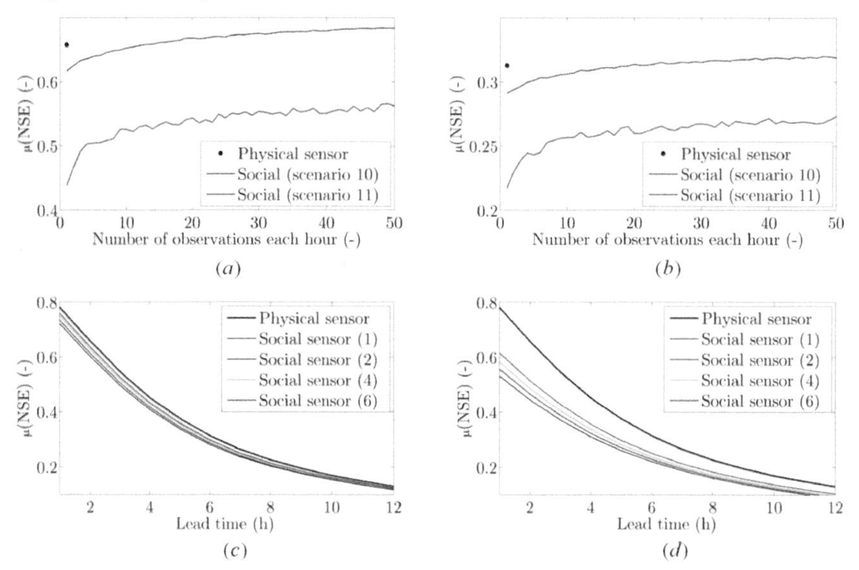

Figure 5.11. Model performance expressed as μ(NSE) values – assimilating observations from physical and social sensors at Leogra gauged station having characteristic described in scenarios 10 (c) and 11 (d)

Figure 5.12 and Figure 5.14 show the results obtained from the Experiment settings represented in Figure 5.3 in case of observations from distributed StPh and StSc sensors. Also in this case, different simulation runs (100) of random values of arrival frequency and uncertainty are performed.

From Figure 5.12, in which observations have the same characteristics of previous scenario 10, it can be seen that the assimilation of observations from the StPh sensor in the Leogra sub-catchment (Setting A) provides a better flood prediction at Ponte degli Angeli if compared to the assimilation of a small number of CS observations provided by a StSc sensor in the same location (Setting B). In particular, Figure 5.12 show that, depending on the flood event, the same NSE values achieved with assimilation of physical observations (hourly frequency and high accuracy) can be obtained by assimilating between 10 and 20 CS observations per hour in case of 4 hours lead time. Such number of CS observations tends to increase for increasing values of lead times. In case of intermittent observations of Figure 5.13, the overall reduction of NSE is such that even with a high number of observations (even higher than 50 per hour) the NSE is always lower than the one obtained assimilating physical observations for any lead time.

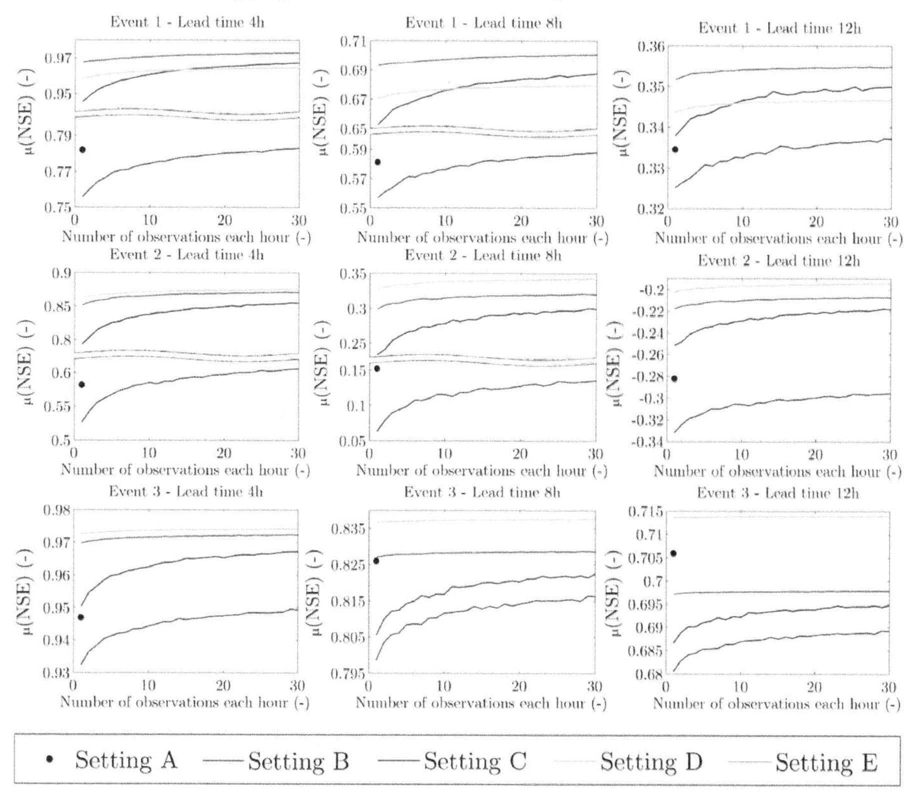

Figure 5.12. Model performance expressed as $\mu(NSE)$ – assimilating different number of crowdsourced observations during the three considered flood events, for the three lead time values, having characteristic of scenario 10

In case of Setting C, it can be observed for all three flood events that distributed StSc sensors in Timonchio, Leogra and Orolo sub-catchments allow to obtain higher model performances than the one achieved with only one StPh sensor (see Figure 5.12). However, in case of flood event 3 this is valid only for low lead time values. In fact, for 8 and 12 hours lead time values, the contribution of CS observations tend to decrease in favor of physical observations from the Leogra sub-catchment. This effect is predominant in case of intermittent crowdsource observations, scenario 11, showed in Figure 5.13. In this case, Setting C has higher μ(NSE) values than Setting A only during flood event 1 and for lead time values equal to 4 and 8 hours.

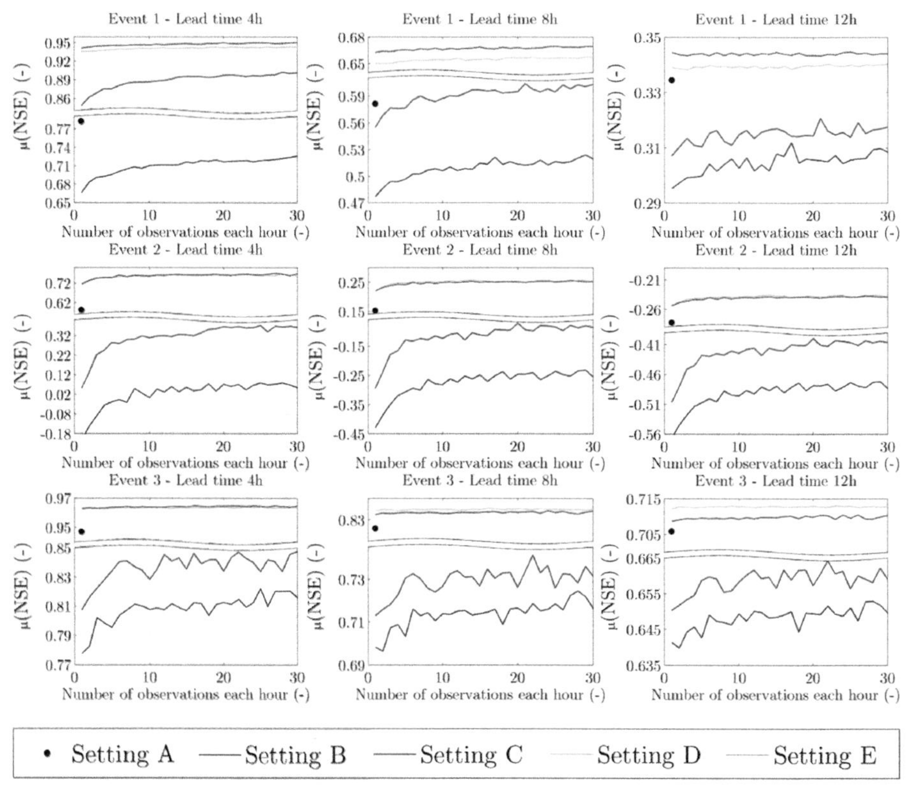

Figure 5.13. Model performance expressed as μ(NSE) – assimilating different number of crowdsourced observations during the three considered flood events, for the three lead time values, having characteristic of scenario 11

It is interesting to note that in case of Setting D, during flood event 1, the μ(NSE) is higher than Setting C for low number of observations. However, with higher number of observations, Setting C is the one providing the best model improvement for low lead time values. In case of intermittent observations (Figure 5.13), it can be noticed that the Setting D provides always higher improvement than Setting C. For flood event 1, the best model improvement is achieved in case of Setting E, i.e. fully integrating StPh sensor with distributed StSc sensors. On the other hand, during flood events 2 and 3 Setting D shows higher improvements than Setting E. In case of intermittent observations the difference between Setting D and E tends to reduce for all the flood events. Overall, settings D and E are the ones which provided the highest μ(NSE) in both scenario 10 and 11. This demonstrates the importance of integrating existing network of StPh sensors (Setting A) with StSc sensors providing CS observations in order to improve flood predictions. As in case of Experiments 1, assimilation of intermittent observations tends to significantly reduce the μ(NSE) values for all settings during the considered flood events.

Figure 5.14 shows the standard deviation of the NSE obtained for the different settings in case of 4 hours lead time. Similar results are obtained for the 3 considered flood events. In case of Setting A, σ(NSE) is equal to zero since observations are coming from StPh sensor at regular time steps. Higher σ(NSE) values are obtained in case of Setting B, while including different CS observations (Setting C) tend to decrease the value of σ(NSE). It can be observed that σ(NSE) decreases for high values of CS observations. As expected, the lowest values of σ(NSE) are achieved including the StPh sensor in the DA procedure (Setting D and E). Similar considerations can be drawn in case of intermittent observations, where higher and more perturbed σ(NSE) values are obtained.

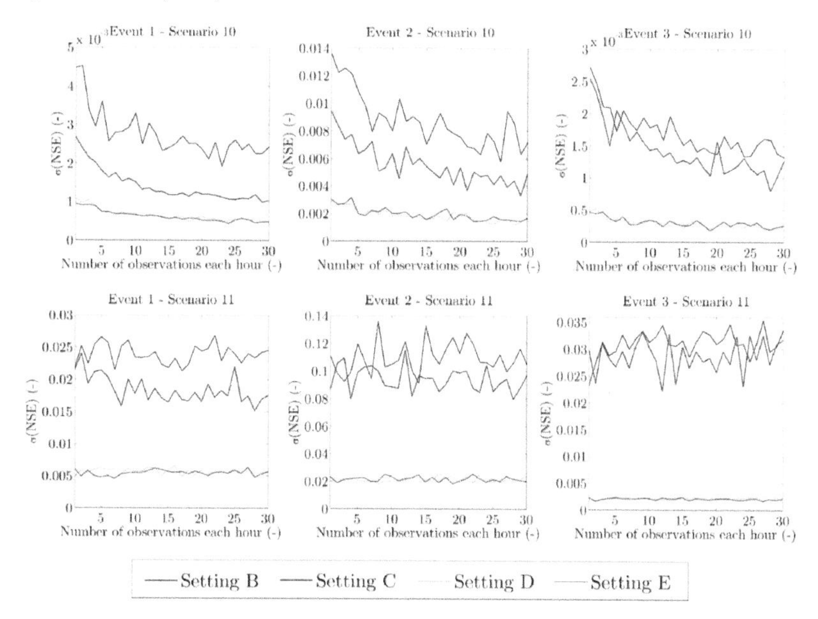

Figure 5.14. Variability of performance expressed as σ(NSE) – assimilating CS observations within setting A, B, C and D, assuming the lead time of 4h, for scenarios 10 and 11

5.5 CONCLUSIONS

This chapter demonstrated that CS observations, asynchronous in time and with variable accuracy, can improve flood prediction if integrated in hydrological models. Such observations are assumed to be collected using low-cost StSc sensors as, for example, staff gauges connected to a QR code on which people can read the water level indication and send the observations via a mobile phone application. In Experiment 5.1 (Brue catchment) the sensitivity of the model results to the different frequencies and accuracies of CS observations coming from a hypothetical StSc sensor at the catchment outlet is assessed. In the Experiment 5.2 (Bacchiglione catchment), the influence of the combined assimilation of CS observations, coming from a distributed network of StSc sensors, and existing streamflow observations from StPh sensors, used in the EWS implemented by AAWA, is evaluated. Due to the fact that CS streamflow observations are not yet available in both case studies, realistic synthetic observations with various characteristics of arrival frequency and accuracy are introduced. Overall, we demonstrated that the results we have obtained are very similar in terms of model behaviour assimilating asynchronous observations in both case studies.

In Experiment 5.1 it is found that increasing the number of CS observations within the observation window increases the model performance even if these observations have irregular arrival frequency and accuracy. It is found that the accuracy of the observations influences the model results more than the actual (irregular) moments in which the streamflow observations are assimilated into the hydrological models. The noise in the NSE is reduced when the assimilated observations are considered having non-periodic behavior. In addition, the intermittent nature of the observations tends to drastically reduce the NSE of the model for different lead times. In fact, if the intervals between the observations are too large then the abundance of CS data at other times and places is no longer able to compensate their intermittency.

Experiment 5.2 showed that, in the Bacchiglione catchment, the integration of observations from StSc sensors and single StPh sensor can improve the flood prediction even in case of a small number of intermittent CS observations. In the case of both StPh and StSc sensors located in the same place the assimilation of CS observations gives the same model improvement as the assimilation of physical observations only when they have high number and non-intermittent behaviour. In particular, the integration of existing physical sensors with a new network of social sensors can improve the model predictions, as shown in the Bacchiglione case study.

Although the cases and models are different, the results obtained are very similar in terms of model behaviour assimilating asynchronous observations.

6

ASSIMILATION OF SYNCHRONOUS DATA IN HYDRAULIC MODELS

In previous chapters, methods to improve hydrological models with crowdsourced information have been presented. However, a similar problem can be formulated for hydraulic models. This chapter aims to assess the effect of data assimilation approaches and sensor positioning in the assimilation of water depth observations from StPh sensors in distributed hydraulic modelling. Lumped and distributed versions of the 3-parameter Muskingum model described in Chapter 2, are implemented in the Trinity and Sabine rivers to study the effect of data assimilation approaches using real-time streamflow observations from StPh sensors in flood predictions. Besides, a Muskingum-Cunge model is applied to a synthetic river with rectangular section and then to a natural river with varying cross-section (the Bacchiglione River, Italy) to assess sensor position effects in the assimilation performance. Multiple synthetic experiments are implemented in both the synthetic and Bacchiglione Rivers to assess impacts of spatial locations of StPh sensors on prediction performance.

This chapter is based on the following peer-reviewed journal publications:

Mazzoleni M., Noh S.J., Lee H., Liu Y., Seo D.J., Alfonso L. and Solomatine D.P. (2016) Real-time assimilation of streamflow observations into a hydrologic routing model: Effects of different model updating methods, Journal of Hydrology, under review

Mazzoleni M., Chacon-Hurtado J., Noh S.J., Alfonso L., Seo D.J. and Solomatine D.P. (2016) Data assimilation in hydrologic routing: impact of sensor placement on flood prediction, Hydrological Processes, Hydrological Processes, under review

6.1 Introduction

DA methods has been increasingly implemented in several applications in order to assimilate physical variables, such as water depth and flood extent, coming from remote sensing or in-situ sensors into hydraulic model in order to improve model performances and consequently reduce uncertainty in flood forecasting (Schumann et al., 2009; Yan et al., 2015).

Over the last couple of years, the assimilation of remotely sensed water level observations in hydrological and hydraulic modelling has become more attracting thanks to their availability and spatially distributed nature (Yan et al., 2015). However, one of the main challenges in assimilating remote sensing observations of flow in hydraulic model is the lack of maturity of processing chains needed to systematically extract these observations (Schumann et al., 2009). The potential of assimilating distributed value of water level from remote sensing for improved discharge and water depth estimation has been explored in different studies (e.g. Andreadis et al., 2007; Neal et al., 2007; Hostache et al., 2010; Matgen et al., 2010; Biancamaria et al., 2011; Giustarini et al., 2011; Mason et al., 2012; García-Pintado et al., 2013; Andreadis and Schumann, 2014) and a detailed review is presented by Schumann et al. (2009) and Yan et al. (2015).

On the other hand, only few studies have implemented state updating techniques on hydraulic and hydrologic routing models using water depth measurements from in-situ sensors (Madsen et al., 2003; Romanowicz et al., 2006; Neal et al., 2007; Ricci et al., 2011; Neal et al., 2012; Kim et al., 2013; Madsen and Skotner, 2005). A procedure based on the Ensemble Kalman Filter is presented in Madsen et al. (2003) in order to assimilate water levels and fluxes observations within the MIKE11 hydraulic model. The available observations are integrated into MIKE11 using assimilation algorithm up to the time of forecast, while after that the model is run in forecast model. Results of such study show significant model improvement assimilating water stage from 3 different gauging stations within the case study. Madsen and Skotner (2005) proposed an adaptive approach based on the combination between a general filtering update combined with error forecasting at measurement points. In the filtering update procedure, a time invariant weighting function is used to distribute model errors from the measurement points to the entire cross sections of the river system, while the aim of the error forecast module is to propagate model errors at measurement points in the forecast period. The results of Madsen and Skotner (2005) showed increasing in the forecast skills for lead times up to 24 hours. Similarly to the previous studies, in Neal et al. (2007) an ensemble Kalman filter algorithm is implemented into a 1D hydraulic model for flow forecasting along the Crouch River using real-time data from 4 sensors. Depending on the sensor location, the assimilation procedure led to an increase in forecast accuracy of between 50% and 70%. In addition, variability in the temporal sampling rate and spatial density of samples had little effect on the accuracy of forecasts. More recently, Neal et al. (2012) proposed an innovative method, based on the Ensemble Transform Kalman Filter, for real-time design of an optimal adaptive space–time sampling of potential measurements from wireless sensor nodes for reducing forecast variance of the flood forecasting. The authors demonstrated that, due to the relatively small distances between network nodes, measurements were highly spatially correlated inducing a limited effect on error variance in case of multiple sensors. The authors suggested a wider application of the method including the assimilation of different data types. Assimilation of water depth observations to a 2-dimensional hydraulic model via particle filtering is proposed by Kim et al. (2013). The authors demonstrated that the methodology contributed to reducing uncertainty in estimation of the Manning's n as well as the rating curve. However, only few studies showed the effect of flow assimilation into hydrologic routing models. In Liu et al. (2008) presented an application of Maximum Likelihood Ensemble Filter (MLEF) method for a hydrologic channel routing model based on the variable three-parameter Muskingum model. Errors in the inflow and outflow observations, and uncertainties in the initial conditions and Muskingum parameters are considered. Similarly, Lee et al. (2011b) applied a 1D-Var method in order to integrate real-time streamflow observations into a 3-parameter Muskingum model. However, no one of the previous studies showed that flood predictions from a simplified hydraulic model can benefits from the assimilation of water depth observations derived from network of randomly distributed sensors.

For this reason, the goals of this chapter are to evaluate the effects of a) different DA approaches on the assimilation of streamflow observations from existing StPh sensors and b) assimilation of water depth observations coming from spatially distributed of StPh sensors into hydraulic modelling. As in case of Experiment 4.1 in Chapter 4, this Chapter does not aim to provide optimal sensor locations.

6.2 METHODOLOGY

Streamflow observations are assimilated within a 3-parameter Muskingum model (described in section 2.4.2, page 50) implemented for the reaches A, B and C of the Trinity and Sabine River to assess the influence of different DA approaches in the assimilation of streamflow observations.

On the other hand, in order to assess the effect of sensor location in the assimilation of WD observation within hydraulic models, using KF, a distributed Muskingum-Cunge model is used along a synthetic river and on the Bacchiglione River. In addition, in case of the synthetic river, impacts of different uncertainty of the boundary and the model are evaluated according to varying locations of observation sensors.

For the Trinity and Sabine Rivers, streamflow observations are coming from existing flow StPh sensors managed by NWS, while for the synthetic and Bacchiglione Rivers, synthetic realistic streamflow observations are used due to the fact that WD observations along the river reaches are not available at the time of this chapter. It is worth noting that, in all the previous case studies, flow and WD observations are considered synchronous and continuous in time.

6.2.1 Data assimilation methods

For the reaches A, B and C along the Trinity and Sabine Rivers, Direct Insertion (DI), Nudging Scheme (NS), Kalman Filter (KF), Ensemble Kalman Filter (EnKF) and Asynchronous Ensemble Kalman Filter (AEnKF) are used (see Chapter 3). In particular, in case of AEnKF, two different values of W equal to 2 and 5 hours are used. On the other hand, for the synthetic and Bacchiglione Rivers, only the KF it is used to assimilate distributed WD observations and improve the model predictions of Muskingum-Cunge model. In particular, the state variables at each reach of the Bacchiglione River (see Figure 2.10) are updated separately.

6.2.2 Observation and model error

Observation error is estimated, as described in the previous chapters, using Eq. (4.4) assuming 0.1 as value of α in case of StPh sensors. On the other hand, the estimation of the model error is more complex in this chapter. In fact, for each case study and DA method, a different characterization of the model error is performed.

Regarding the Trinity and Sabine Rivers, different DA approaches are implemented into the 3-parameter Muskingum models. In all reaches A, B and C, the implementation of DI does not request any estimation of the model error since it is assumed that observations are perfect and thus the model state is replaced with such observations. In case of reaches A and B (lumped model), for NS and KF, the covariance matrix **S** it is initially estimated as diagonal matrix ($n_{state} \times n_{state}$) having as elements the variance between observed and simulated streamflow, i.e. the state of the Muskingum-Cunge model. However, in order to assess the effect of assimilation of streamflow observations, as it is performed in Chapter 5, the model is assumed more accurate than the observations. That is why, the elements of the covariance matrix S are assumed equal to 90 m^6/s^2. In case of EnKF and AEnKF, both boundary and parameters are perturbed following Eqs.(4.2) and (4.3) in order to analyses the sensitivity of the lumped model and effectively represent model error. In particular, three different values of the perturbation factors ε_I and ε_P are used. The results of such analysis are showed in Figure 6.1.

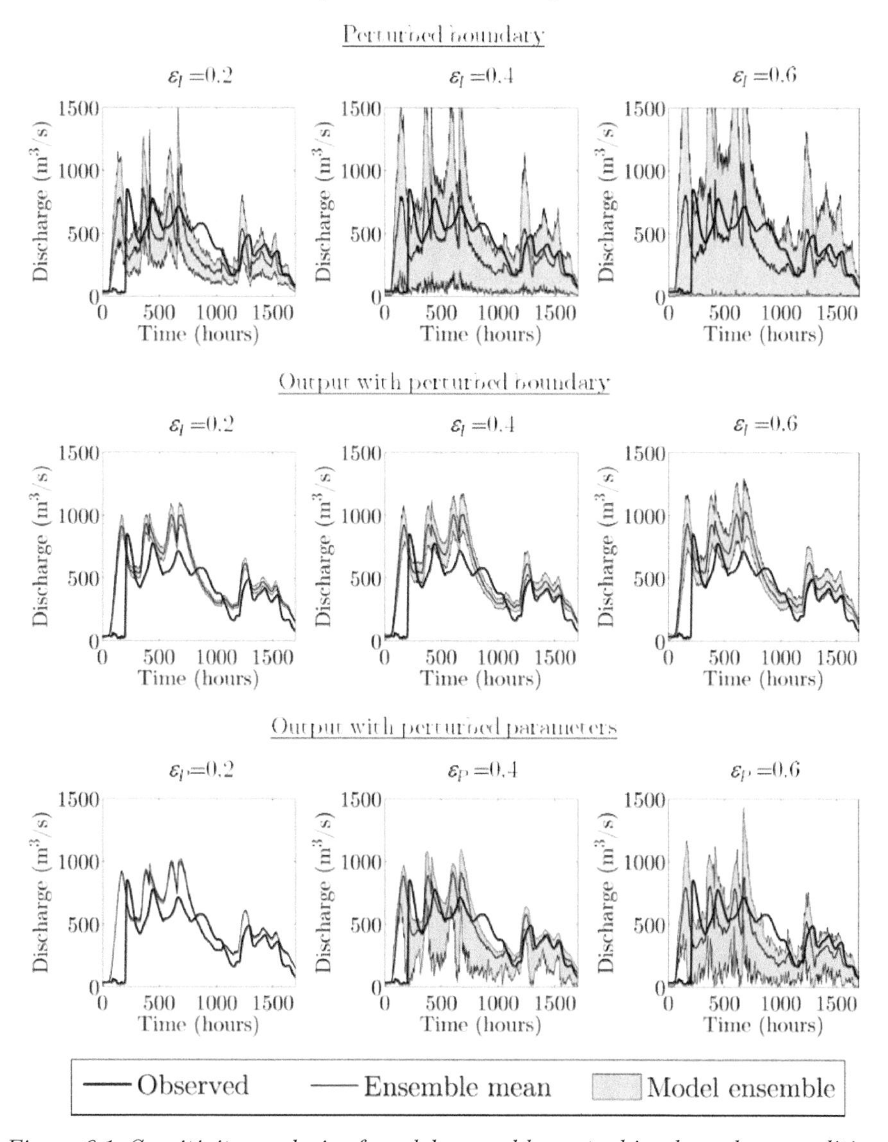

Figure 6.1. Sensitivity analysis of model ensemble perturbing boundary conditions or model parameter with different perturbation factor values during event A.1

It can be observed that perturbing the boundary conditions (model input) of the lumped 3-parameter Muskingum model, without update, induced a reduction of such initial spread (first row in Figure 6.1) in the outlet section (second row in Figure 6.1) during flood event A.1. In fact, in order to have a proper model spread at the river outlet, a high value of $_{I}$, e.g. 1.5 times the value of the boundary itself. However, such value of $_{I}$ might not provide physical meaning to the definition of the ensemble spread. This can be related to the structure of the hydraulic model itself. On the other hand, when the model parameter , K and X are perturbed, the spread at the outlet section is high also in case of low value of $_{P}$.

For this reason, in case of lumped model, the model ensemble is generated perturbing the model parameters in such a way that the variance of the ensemble spread is equal to the value of **S**, estimated for NS and KF, for a value of ε_P equal to 0.25 for reach A and 0.18 for reach B. In this way, it can be possible to later compare the results achieved assimilating streamflow observations using NS, KF, EnKF and AEnKF since the model error is univocally defined. A value of N_{ens} equal to 50 is considered in this chapter. In case of reach C, i.e. distributed model, the same value of S used for reaches A and B (the lumped model) is assumed in case for NS and KF. Such value is considered stationary and constant along the river reach. Regarding the definition of the model error for EnKF and AEnKF, it is found that perturbing model parameters in case of distributed structure might induce instability in the hydraulic model itself. That is why, the model input is perturbed considering ε_I equal to 0.35 in order to achieve a variance of the ensemble spread of 90 m^6/s^2., as in case of lumped model. Also in this case, a value of the value of N_{ens} equal to 50 is considered. It is worth noting that in both lumped and distributed model implemented in the Trinity and Sabine Rivers, the covariance error $\mathbf{M_b}$ is assumed equal to 0 in case of KF. In Table 6.1, a summary of the model error definition in the different case studies and DA approaches is reported.

Table 6.1 Definition of the model error in case of different river reaches and DA methods

	NS – KF	EnKF	AEnKF
	S (m^6/s^2)	ε_I (-)	ε_P(-)
Reach A (lumped)	90	-	0.25
Reach B (lumped)	90	-	0.18
Reach C (distributed)	90	0.35	-

In contrast to the previous two rivers, in the synthetic and Bacchiglione Rivers only a KF is used to assimilate WD at different location along the reaches. In case of the synthetic river, knowing the observed and simulated time series of the upstream boundary conditions it is possible to estimate the boundary covariance matrix $\mathbf{M_b}$ in both the considered flood events. The model covariance matrix $\mathbf{S}$ is estimated, at each time step, as function of model state $\mathbf{x}$ at time t, without any model update, in each cross as:

$$\mathbf{S}_t^j = \left(\varepsilon_m \cdot \mathbf{x}_t^j\right)^2 \tag{6.1}$$

In order to assess the effect of model and boundary conditions error in the DA performances, three different scenarios are considered and explained in the next section. The coefficient ε_m is defined in the next section, where different scenarios of boundary and model error are introduced.

For the Bacchiglione River, the boundary conditions errors $\mathbf{M_b}$ for Reaches 1, 2 and 5 in the headwater catchments (see Figure 2.10) are calculated by comparing the observed and simulated hydrographs derived using the hydrological model. For Reaches 3, 4 and 6, $\mathbf{M_b}$ consists of the error in the flow from the upstream reaches and that from the inter-basin estimated with the hydrological model. Figure 6.2 shows that $\mathbf{M_b}$ in Reach 3 ($\mathbf{M_{b,3}}$) is a function of the error covariance matrix of the flows at the outlets of Reaches 1 and 2 ($\mathbf{M_{out,1}}$ **and** $\mathbf{M_{out,2}}$) and the flow in I_3, $\mathbf{M_{I3}}$. In case of reach 3, we assumed that the boundary condition error, $\mathbf{M_b}$, in Reaches 1 and 2 is larger than the model error $\mathbf{M_{m,1}}$ and $\mathbf{M_{m,2}}$.

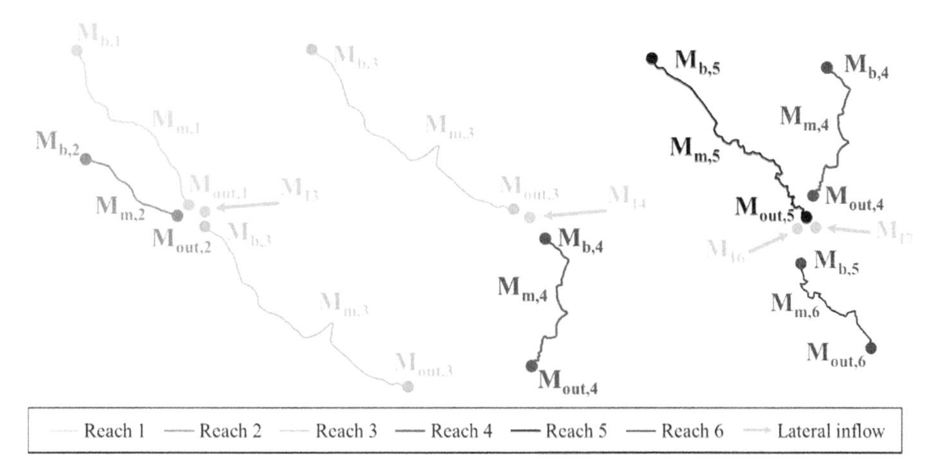

Figure 6.2. Representation of the errors estimation in Bacchiglione catchment based on the distribution of its reaches

This means that we can assume $\mathbf{M_{out,1}}$ **and** $\mathbf{M_{out,2}}$ to have the same magnitude as $\mathbf{M_{b,1}}$ and $\mathbf{M_{b,2}}$. At this point, it is possible to estimate $\mathbf{M_{b,3}}$ as maximum values between $\mathbf{M_{out,1}}$, $\mathbf{M_{out,2}}$ and $\mathbf{M_{I3}}$. The same procedure is followed for the reach 4 and 6 as reported in Figure 6.2. However, in case of available observed streamflow values at the reach outlet, $\mathbf{M}_{out}$ should be estimated as a function between observed and simulated values in the same way it is done with reach 1, 2 and 5. In case of flood event 1, the values of $\mathbf{M}_{b,1}$, $\mathbf{M}_{b,2}$, and $\mathbf{M}_{b,5}$, are set equal to 63.3m^6/s^2, 2316.4 m^6/s^2 and 214.2 m^6/s^2, respectively. In addition, **S** is set 39.7 m^6/s^2 in each river reach comparing the simulated streamflow with Muskingum-Cunge and MIKE11 model, using this last one as benchmark.

6.2.3 Streamflow observations

In case of Trinity and Sabine Rivers, the observed streamflow observations available at the NWS stations, described in Section 2.4.2 (page 52), are used.

On the other hand, for the synthetic river, the observed realistic WD values along the reach are generated using the time series of recorded streamflow values at the outlet of the Brue catchment as perfect boundary condition for the MC model. The conceptual lumped model used in Chapter 5, is used to estimate the upstream boundary that led to the estimation of the simulated WD values. The average rainfall, used as input in the hydrological model, is estimated using the Ordinary Kriging, in order to optimal interpolate the point data from the 49 rainfall station information available in the Brue catchment (Matheron, 1963).

Regarding the Bacchiglione River, synthetic water depth observations are generated using the same method described in Section 5.2.3 (page 112). In fact, using the observed time series of precipitation, the output from the hydrological model are used as input in the Muskingum-Cunge model to then estimate the realistic synthetic water depth observations for each x (1000m) along the Bacchiglione River. Forecasted input are used to estimate the simulated water depth along the river reach.

6.3 EXPERIMENTAL SETUP

6.3.1 Experiment 6.1: Effect of different DA methods

In reaches A and B (Experiment 6.1.1), a lumped version of the 3-parameter Muskingum-Model is used to assimilate streamflow observations during 6 flood events (A.1, A.2, A.3, B.1, B.2 and B.3) at the StPh station of TDDT2 and DWYT2 respectively. The description of such flood events is reported in section 2.4.2 (page 50). Both simulated and forecasted hydrograph at the outlet sections of reaches A and B are compared to the observed value and the performance measures NSE, R and Bias are then estimated.

Regarding reach C (Experiment 6.1.2), a distributed structure of the 3-parameter Muskingum model is implemented to assimilated streamflow observations. In particular, three different situations of assimilation location are introduced: a) assimilation of streamflow observations at the StPh sensor of Grand Praire (GPRT2), b) assimilation of streamflow observations at the Dallas StPh sensor (DALT2), and c) assimilation of observations coming from both GPRT2 and DALT2 StPh sensors. Simulated and observed streamflow values are compared to the observed ones at DALT2 because of the strategic position of such StPh sensor in terms of flood risk managmente for the Dallas area.

6.3.2 Experiment 6.2: Effect of sensors location on KF performances

In Experiment 6.2, flow propagation and consequent assimilation of distributed WD observations are performed in the synthetic (Experiment 6.2.1) and Bacchiglione Rivers (Experiment 6.2.2) using a Muskingum-Cunge distributed model. It is worth noting that the flood events described in Section 2.3.2 (page 45) occurred in 2013, 2014 and 2016 are used in the following analysis of the Bacchiglione cathcment.

In Experiment 6.2.1, different values of model error are assumed, see Eq.(6.1*)*. In the following analyses, 3 different scenarios of model and boundary errors are considered. Due to the fact that boundary condition error is assumed fixed, ε_m is changed in order to meet the conditions described in the three scenarios.

- Scenario 1: $\overline{\mathbf{S}} \approx \mathbf{M_b}$, where $\overline{\mathbf{S}}$ is the average value of **S** in time and space. The value of ε_m is set to 0.35 for both flood events
- Scenario 2: $\overline{\mathbf{S}} >> \mathbf{M_b}$. In this way the Muskingum-Cunge model is considered as main source of error in the flood propagation ($\varepsilon_m = 0.8$);
- Scenario 3: $\overline{\mathbf{S}} << \mathbf{M_b}$. In this way model error **S** is considered negligible with respect to boundary error ($\varepsilon_m = 0.01$).

In these scenarios, assimilation of WD at one given location is considered. In Table 6.2, the value of $\overline{\mathbf{M}_m}$ and $\mathbf{M_b}$ are reported for both flood events.

Table 6.2. Value of boundary and model errors $\boldsymbol{M_b}$ and $\boldsymbol{M_m}$ according to the different experimental scenarios and flood events

	$\mathbf{M_b}$	$\overline{\mathbf{M}_m}$ Scenario 1	$\overline{\mathbf{M}_m}$ Scenario 2	$\overline{\mathbf{M}_m}$ Scenario 3
Event 1	41.54	43.64	228.01	0.04
Event 2	31.03	37.73	197.13	0.03

The focus of these Experiment 6.2.2 is to understand how assimilation of distributed WD observations coming from hypothetical StPh sensors can affect the improvements of the MC model at the outlet point of Vicenza. This series of experiments is divided in synthetic and real-world experiments. In the synthetic experiments, different hypothetical locations of StPh sensors along the 6 river reaches of the Bacchiglione River are considered in order to study the sensitivity of model results to sensor positioning during three different flood event. In this case, simulated results are compared with synthetic observed WD value at the outlet section of Ponte degli Angeli (Vicenza).

On the other hand, in real-world experiments observed observations are assimilated at the existing StPh sensors located at PM and PA as reported in Figure 2.10. Different lead time values are used in order to evaluate the predictive capability of the Muskingum-Cunge model assimilating WD observations at different locations. The updating frequency is considered equal to the observation interval which is 1 hour.

6.4 Results and discussions

6.4.1 Experiment 6.1

Experiment 6.1.1: Reach A and B (Trinity and Sabine Rivers)

In Experiment 6.1.1, assimilation of streamflow observations using DI, NS, KF, EnKF, AEnKF (with W equal to 2 and 5 hours) in applied in a lumped 3-parameter Muskingum model along reaches A and B. Figure 6.3 shows the observed and simulated hydrograph, with or without update, in case of reach A (first row) and reach B (second row) during 6 flood events. It can be seen that DI is the DA method that provides better model improvements in all the considered flood events. That it can be due to the lumped structures of the model and to the fact that observations, from StPh sensors, are assumed perfect if compared to the hydraulic model.

On the other hand, assimilation of streamflow observations using NS produces the worst results among all the others DA approaches. AEnKF tends to perform better than KF and EnKF, in particular increasing the value of W, i.e. the number of past observations. In fact, AEnKF with W equal to 5 gives comparable results than DI. In this chapter, the maximum value of W is set to 5 due to the fact that additional past observations would not add any further improvement to the model results. It is interesting to notice that KF and EnKF provide different model results even if the model error is defined in a consistent way. These results show how model error estimation in Kalman filtering can significantly affect the DA performances.

In particular, the bad model performances achieved with EnKF might be due to a low spread of the model ensemble, meaning that the DA method trusts the model more than the assimilated observations. A better definition of the ensemble spread, using for example Eqs.(3.17), (3.18), (3.19), (3.20) and (3.21) in section 3.5 would give higher trust to the streamflow observations and consequent better model results. However, in this chapter the error estimation of the EnKF is defined in such a way to be consistent with the one of KF.

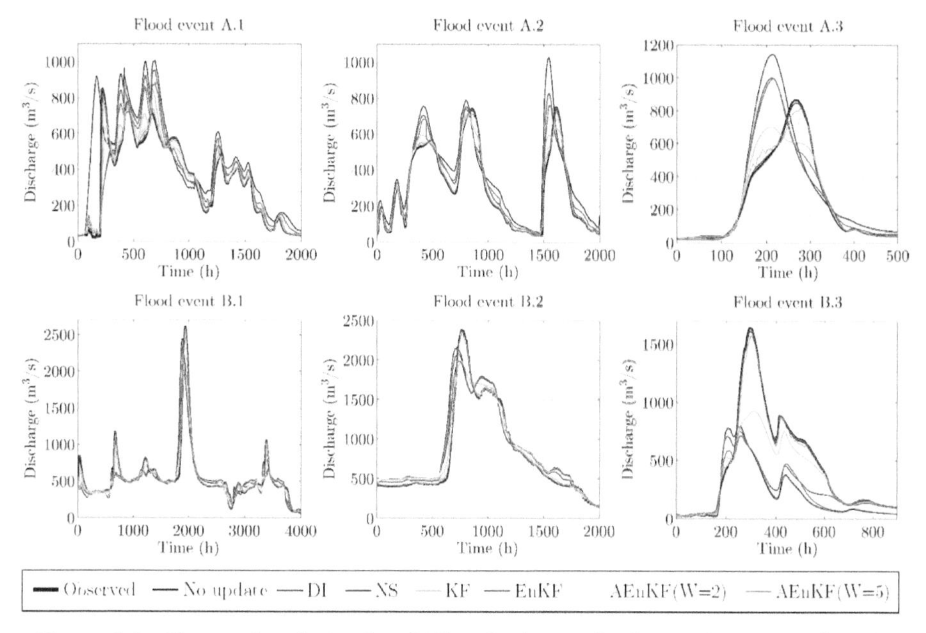

Figure 6.3. Observed and simulated Flow hydrograph obtained in reach A and B using different DA methods

In Figure 6.4 and Figure 6.5, three different performance measures, NSE, R and Bias, are reported for reach A and B during the 6 different flood events. The results reported in Figure 6.4 and Figure 6.5 pointed out a similar trend of NSE, R and Bias in all the flood events in both reaches. In fact, lower value of NSE are obtained without any model update. Best model improvements, in terms of NSE, are achieved with DI and AEnKF, while NS and EnKF are the DA methods which affects the less model results. As expected, AEnKF with W=5 performs better than AEnKF with W=2 due to the higher number of past observations included in the DA procedure. EnKF tends to provide better model improvement than NS in all the considered flood events. Overall, high correlation values are achieved assimilating streamflow observations using all the DA methods. This can be due to the fact that the lumped structure of the model. The minimum R value of 0.8 (flood event A.3) is obtained with NS.

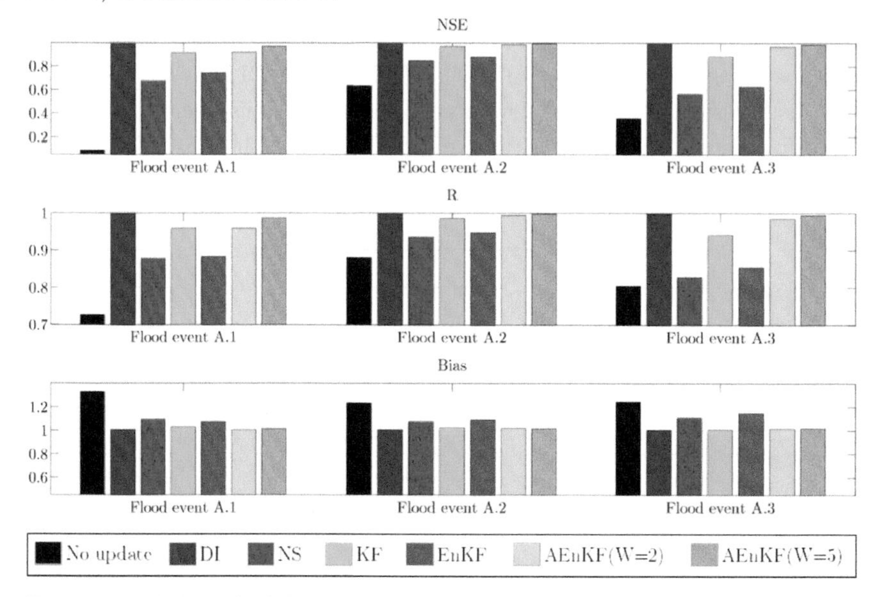

Figure 6.4. Value of NSE, R and Bias obtained using different DA methods in case of reach A during three different flood events

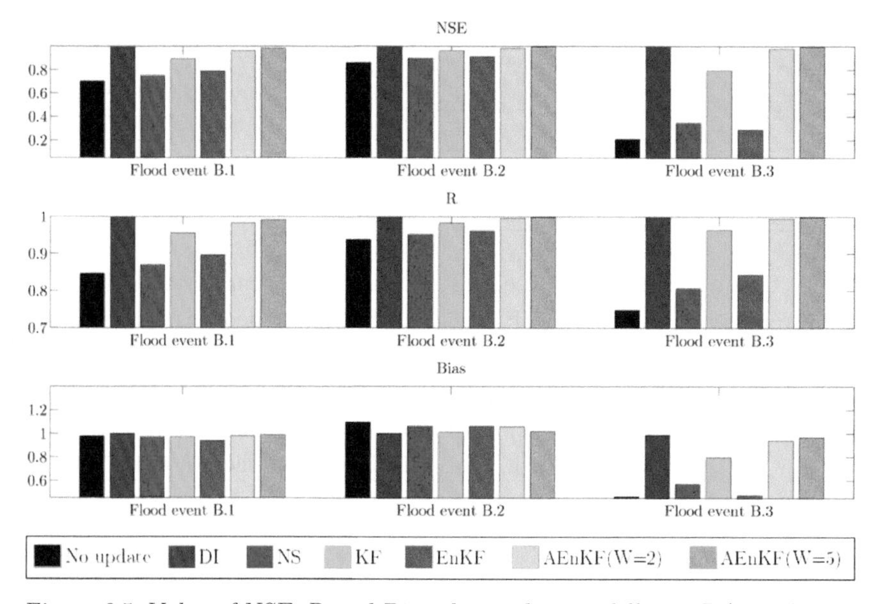

Figure 6.5. Value of NSE, R and Bias obtained using different DA methods in case of reach B during three different flood events

In particular, the highest R values are obtained implementing DI, while comparable results between KF and AEnKF with W=2 can be observed. Bias values close to 1 are achieved with DI and AEnKF with W=5, while the highest and lowest Bias value of 1.23 and 0.53 are obtained in case of EnKF for flood event B.3 and A.3, respectively. Such extreme values of Bias can be related to the ensemble definition of EnKF. Regarding the Bias, it can be see that, overall, streamflow in reach A is overestimated (Bias>1) while opposite results are obtained in reach B where observed streamflow is underestimate (Bias<1). This can be due to an uncertain calibration of the parameter α, see Eq.(2.29), which represent lateral inflow along the river reach.

In Figure 6.6, the Taylor diagrams (Taylor, 2001) for reach A and B during the six flood events are represented. Taylor diagrams are commonly used to graphically summarize how closely simulations fit observations. Such similarity is calculated by means of three statistic as root mean square difference (RMSD), correlation and standard deviation between observations and simulations. This means that, in Figure 6.6 the closest is DA method to the observations (black point) the better. Similar conclusions to the ones just mentioned can be summarized in Figure 6.6.

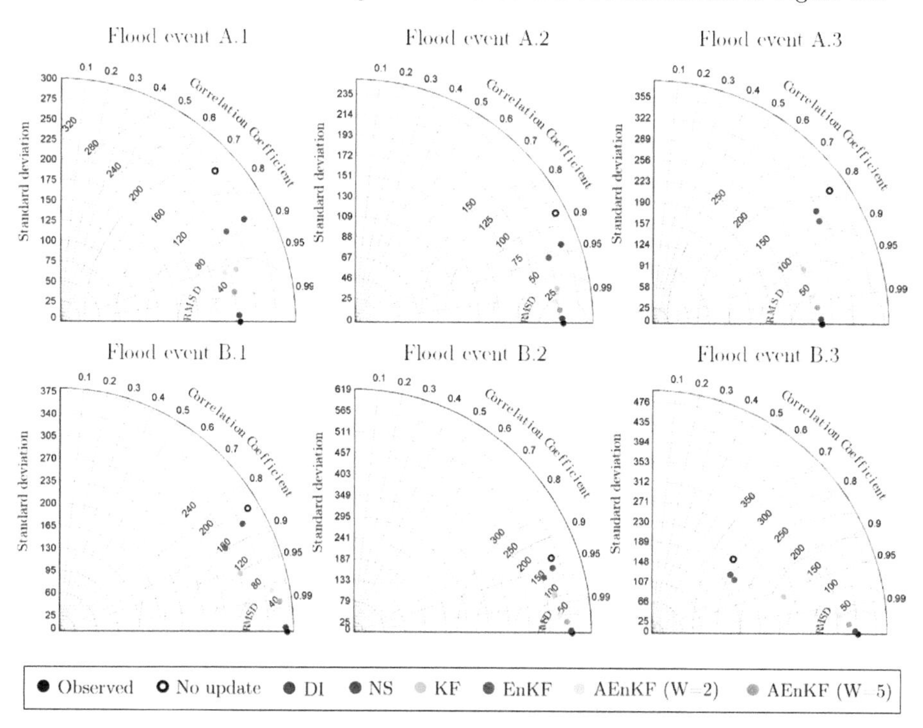

Figure 6.6. Taylor diagrams obtained in reach A and B using different DA methods

The comparison between model predictions in case of different lead time values are showed in Figure 6.7 during flood events A.1 and B.2. As mentioned before, DI provides the best model improvement also in case of different lead time values during both flood events. Good model improvement are also achieved, as previously showed, using AEnKF and KF.

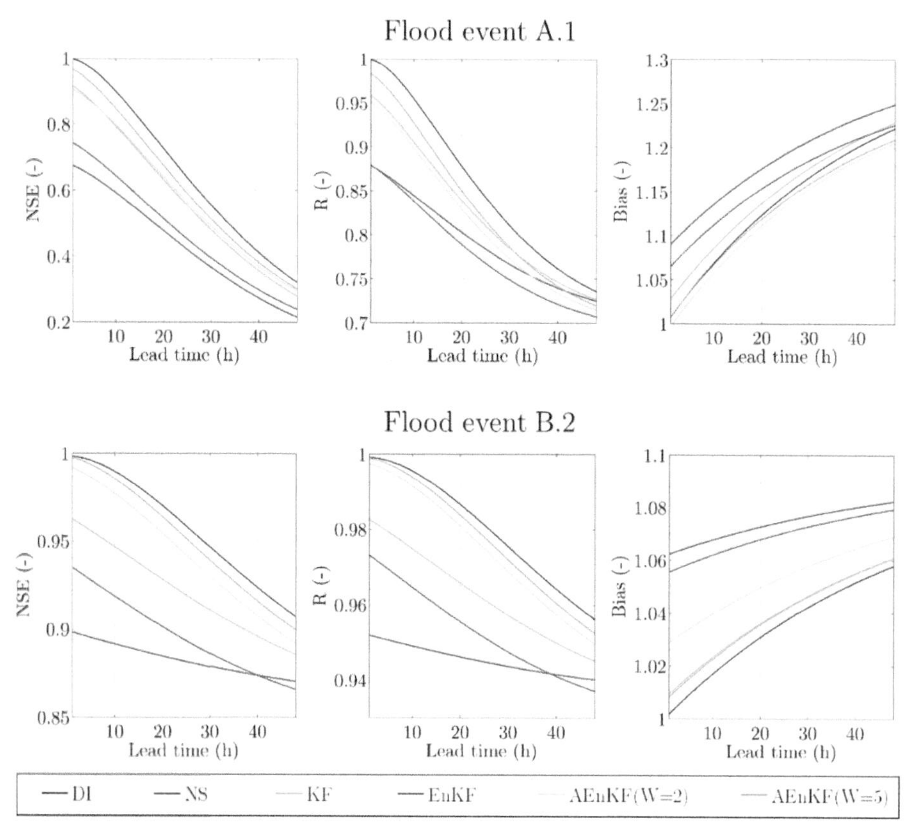

Figure 6.7. Comparison between observations and model predictions, in terms of NSE, R and Bias, in case of different lead time values during flood events A.2 and B.1

However, in case of flood event A.1, NS tends provide higher value of R and lower of Bias if compared to KF, and AEnKF for lead time higher than 35 hours. Moreover, it can be seen that NS, in case of flood event B.2, gives lower NSE and R values than EnKF for low lead time values, while, after 40h lead time, EnKF tends to underperform if compared to NS. However, this behaviour it is not observed in case of Bias, where EnKF provides better results of NS for any lead time value.

Experiment 6.1.2: Reach C (Trinity River)

Regarding Experiment 6.1.2, a distributed 3-parameter Muskingum model is used to estimate streamflow values each 1000m along the Trinity River flowing from Fort-Worth until Dallas. Streamflow observations are assimilated from 2 StPh sensors located in Grand Prairé (GPRT2) and Dallas (DALT2) during the May-June flood event occurred in the DFW area.

Interesting results are showed in Figure 6.8. In fact, in case of distributed model, DI does not provide the best model improvement at the Dallas section if compared to the other DA methods. In fact, assimilation of streamflow observations, e.g. in GPRT2, updates the model states only at that particular location while upstream the assimilation point there are no changes due to the diffusive nature of the Muskingum model. Similar consideration is valid also in case of NS.

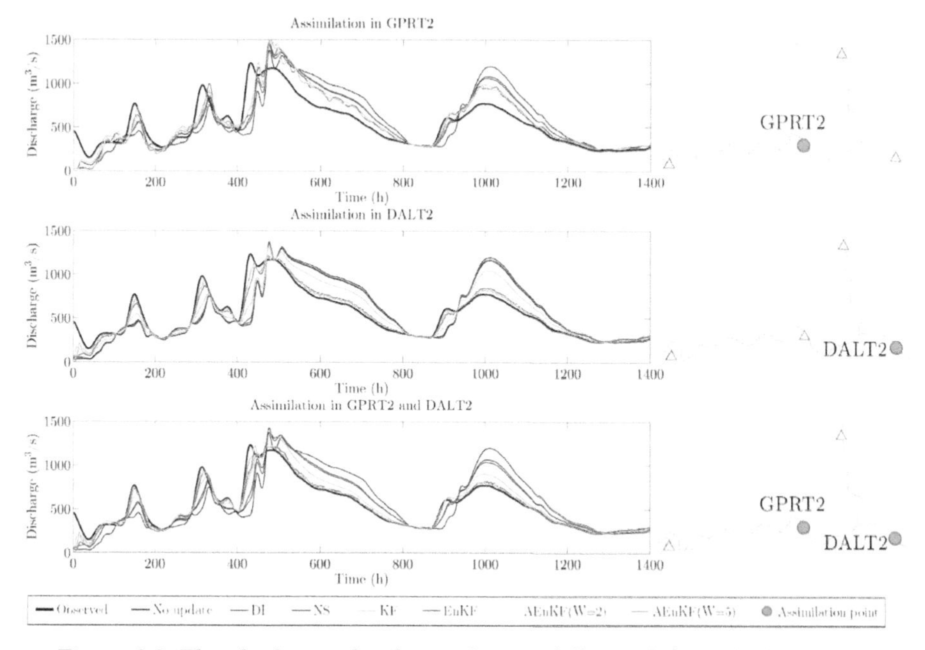

Figure 6.8. Flow hydrographs obtained using different DA methods in case of different location of the StPh sensors

On the other hand, on case of Kalman filtering methods, a higher model improvement is achieved. AEnKF with W equal to 2 and 5 gives very similar model results, while EnKF performs better than KF in all three scenarios of location of the assimilation point. Assimilating streamflow observations from the StPh sensor located at DALT2 insures better DA results than assimilation in GPRT2. This can be due to the influence of sub-reach C.2 in the streamflow estimation in DALT2.

The values of NSE, R and Bias, obtained in case of different location of the StPh sensors, are reported in Figure 6.9. High NSE values are achieved in case of AEnKF with W=5, followed by AEnKF with W=2. DI and NS gives lowest NSE values. Also in case of distributed model, the model error is defined consistently for both KF and EnKF. However, the comparison between NSE values show different results using KF and EnKF, highlighting that, also in this case, Kalman filtering methods are very sensitive to the proper model definition. An increasing trend of the R values it can be seen passing from DI to AEnKF. Overall, high R values are achieved even in case of no model update. In all the three scenarios of sensors location, streamflow in DALT2 is always overestimated (Bias>1), with a minimum and maximum values of 1.002 and 1.12 obtained in case of assimilation in DALT2 with AEnKF (W=5) and assimilation in GPRT2 with DI, respectively. The regular overestimation of the observed flow can be due to an uncertain calibration of the parameter α, see Eq.(2.29), as showed in case of Lumped model.

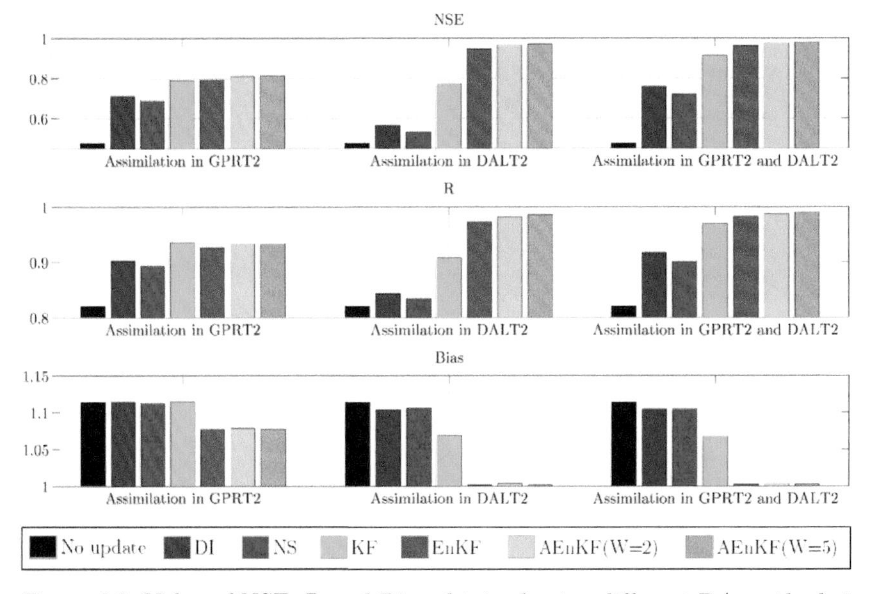

Figure 6.9. Value of NSE, R and Bias obtained using different DA methods in case of different location of the StPh sensors

In Figure 6.10, the Taylor diagrams for different location of the StPh sensor are represented. The Taylor diagrams show that, as expected, lower standard deviation and RMSD are achieved in case of assimilation in both GPRT2 and DALT2. In particular, good model improvements are obtained with the StPh sensor in DALT2, with a significant improvement of EnKF and AEnKF, while opposite results can be seen in case of assimilation only in GPRT2. This suggests that additional sensors in close to the Dallas area would improve further model results, helping in reduce flood impact and consequent damages.

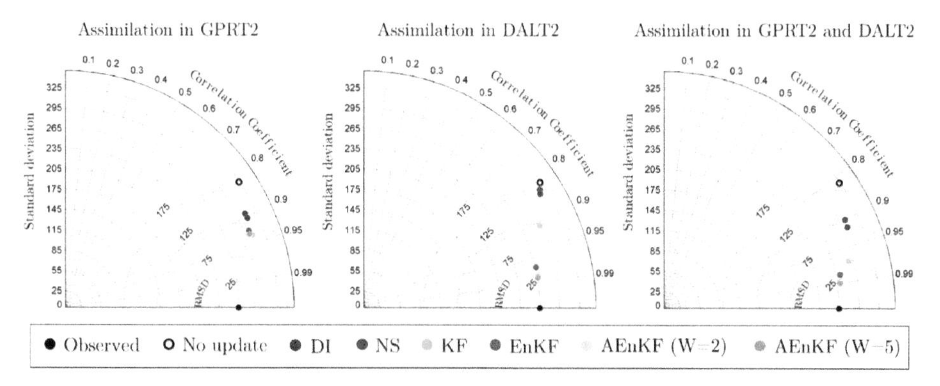

Figure 6.10. Taylor diagrams obtained using different DA methods in case of different location of the StPh sensors

Overall, NS is the method that provides worst model improvement, as in case of lumped model, while AEnKF is best DA methods, among the one used in this chapter. Increasing the number of past observations included in the AEnKF helps to increase model performances. Figure 6.11 shows the streamflow profile along sub-reach C1 obtained in case of assimilation in GPRT2 at three different time steps.

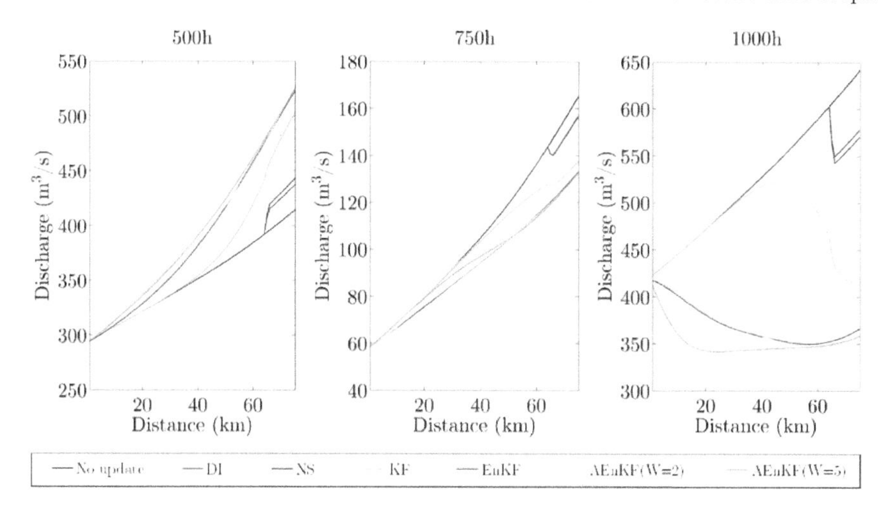

Figure 6.11. Streamflow profile along the sub-reach C1 at three particular time steps in case of assimilation in GPRT2

As previously discussed, both assimilation using DI and NS does not affect the profile upstream the assimilation point. Different results are obtained in case of

Kalman filter methods. In such methods all the model states are updated as response of assimilation in only one location. This is due to the distributed nature of the Kalman gain matrix **K**. In fact, as it is showed in the next section, the maximum value of **K** is achieved at the assimilation point, while a reduction of such value can be found more we move away from the assimilation point. Another interesting aspect is that, in case of KF, a discontinuity in the profile can be observed at the assimilation point. However, such discontinuity is not present in case of ensemble methods.

The same behaviour can be observed in Figure 6.12 where the water storage (m^3) between one cross section and the previous one is calculated along the sub-reach C1 for the same time steps used in Figure 6.11. In fact, DI, NS and KF induce an abrupt change in the water storage at the sensor location, while in ensemble methods a smooth variation of storage is observed along the river. However, an interesting fact it can be observed in both Figure 6.11 and Figure 6.12. In ensemble methods, the states update upstream the assimilation point induces a non-univocal profile in both EnKF and AEnKF. This is due to the ensemble generation at the boundary conditions which generates different ensemble trajectories upstream the assimilation point. In the first row of Figure 6.13 the ensemble of streamflow profiles is represented for both EnKF and AEnKF. It can be observed that AEnKF provides lower ensemble than EnKF. In addition, the ensemble spread, which is a function of the model error, is reducing at the sensor location, as expected. One possible solution to reduce the randomness of the streamflow profile upstream the sensor location is to assume a different location of such sensor.

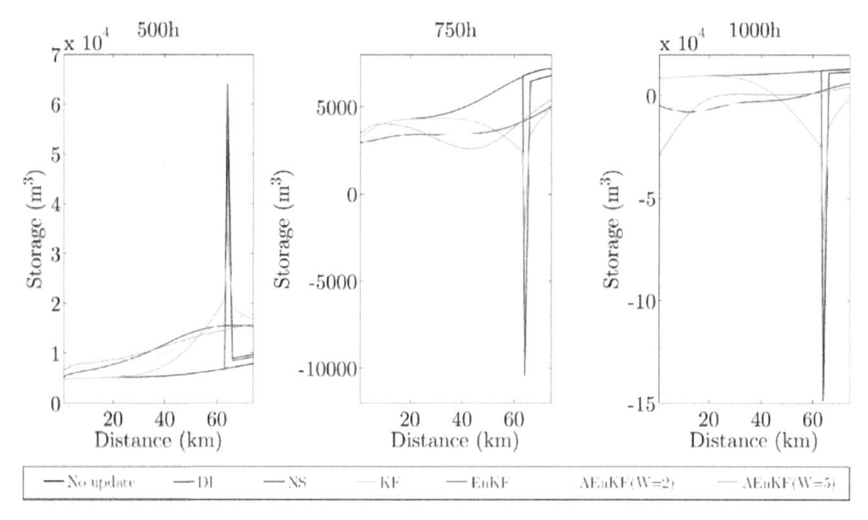

Figure 6.12. Water storage profile along the sub-reach C1 at three particular time steps in case of assimilation in GPRT2

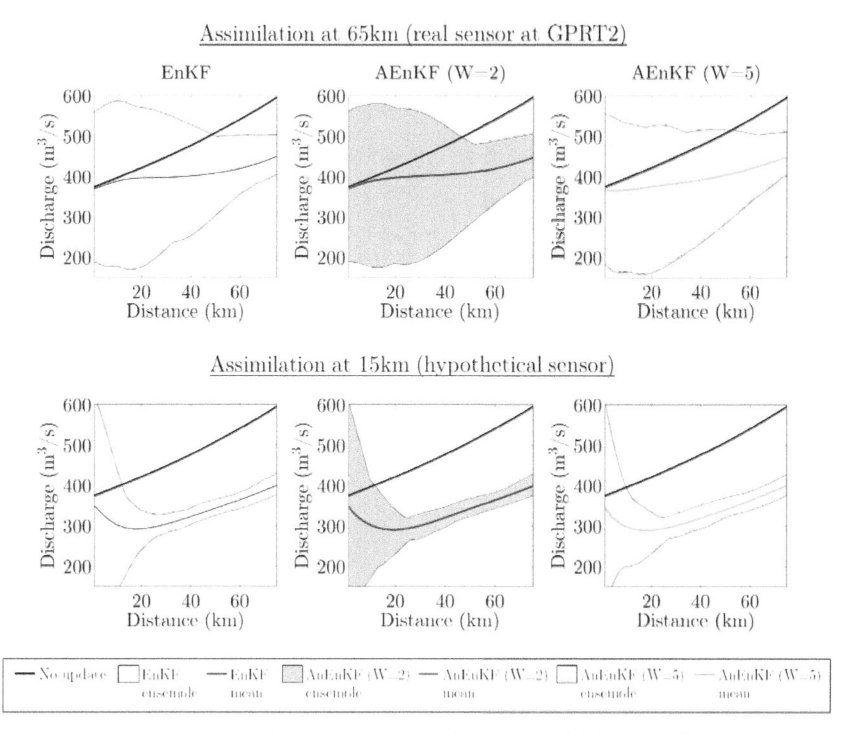

Figure 6.13. Ensemble of streamflow profiles using EnKF and AEnKF at time 1000h, in case of two scenario of StPh sensor GPRT2 location

For this reason, in the second row of Figure 6.13 the model results obtained locating the sensors upstream it is represented, and exactly at 15km from the upstream boundary of sub-reach C1. In this case it can be demonstrated that model error is significantly reduced at this hypothetical StPh sensor location, inducing a lower error downstream such point. Locating StPh sensor upstream the river reach helps to reduce the uncertainty in the profile estimation upstream the assimilation point. This can be particularly useful for flood damage reduction purposes.

In Figure 6.14, the NSE, R and Bias values of model prediction, up to 24 hours lead time, are showed in case of peak1. It can be noticed that assimilation in DALT2 provides an overall improvement of the model predictions. However, such improvement is lost after few hours, leading to NSE, R and Bias values equal to the ones obtained without any model update. On the other hand, assimilation in GPRT2 gives higher value of the statistical indexes in case of high lead time values. This is due to the propagation effect from the assimilation point up to the target point in DALT2. As described before, AEnKF and EnKF provides better model performances for any lead time values, while DI and NS are the less effective DA methods in case of distributed model in all scenarios of sensors location.

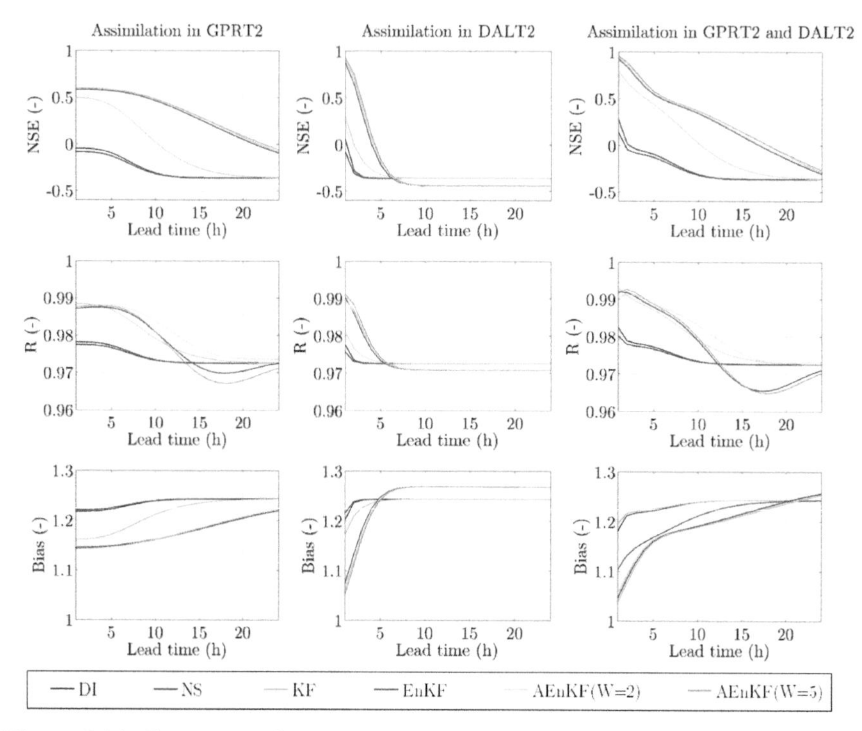

Figure 6.14. Comparison between observations and model predictions, in terms of NSE, R and Bias, in case of different lead time values and different StPh sensors location during peak 1 in the May-June 2015 flood event

6.4.2 Experiment 6.2

Experiment 6.2.1: Synthetic River

Figure 6.15 shows the model improvements at two particular river section assimilating WD observations at the assimilation point (AP) of 15km of the synthetic river during the two considered flood events in case of Scenario 1. From the analysis of the results it can be noticed that assimilation of WD at a specific location it has an impact along the whole river reach and not only at the AP. This can be related to the distributed structure of the MC model and the propagation effect intrinsic in the MC. In addition, the way KF updates model states might be an additional reason to motivate these results. For this reason, in Figure 6.16 the difference between observed and simulated model states (first row), Kalman gain **K** (second row) and model error covariance matrix **P** (third row) during flood event

A are represented, in case of Scenario 1, and three different APs at 15km, 30km and 45km.

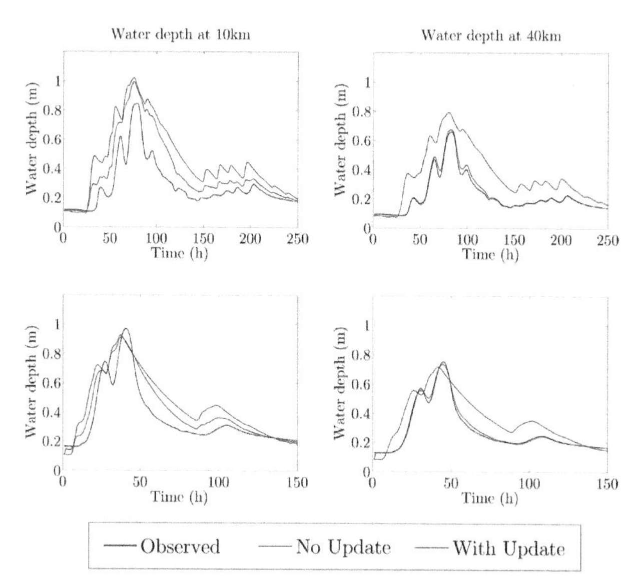

Figure 6.15. Water depth hydrographs obtained at 2 river sections during flood event A and B assimilating observed water depth at the 15km

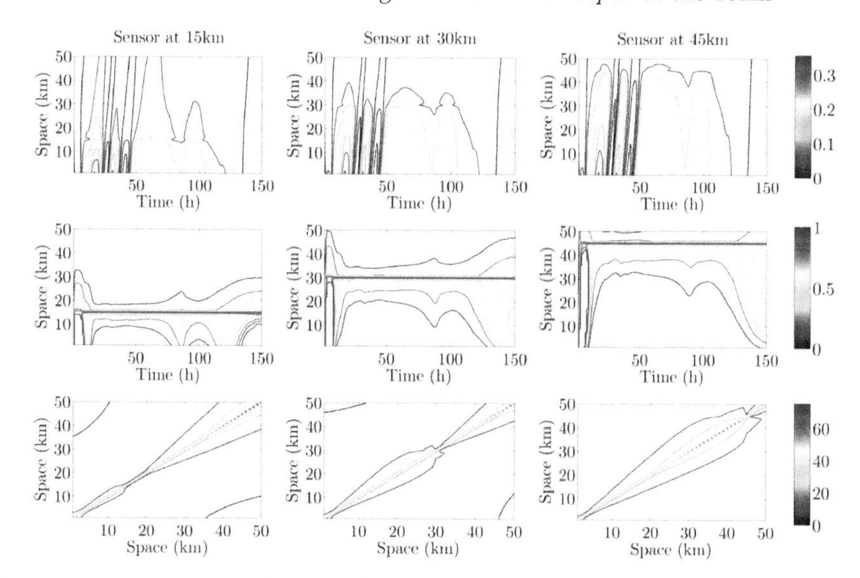

Figure 6.16. First row: difference between observed and simulated (with update) model states; Second row: Kalman gain K (-); Third row: Model error covariance matrix P (m6/s2) at 100 hours. All the previous graphs are referred to event A

As it can be seen, difference between observed and simulated WD is higher upstream the AP than downstream. The maximum value of the Kalman gain **K** is achieved at the AP, as expected. Overall, KF updating effects tends to be propagated both upstream and downstream. However, due to the fact that the Muskingum-Cunge model does not account for backwater effect, such updating effect affects more the locations downstream the AP than upstream. Figure 6.16 shows that the symmetric model covariance matrix **P**, obtained at model time step 100h, has its smallest values at the AP.

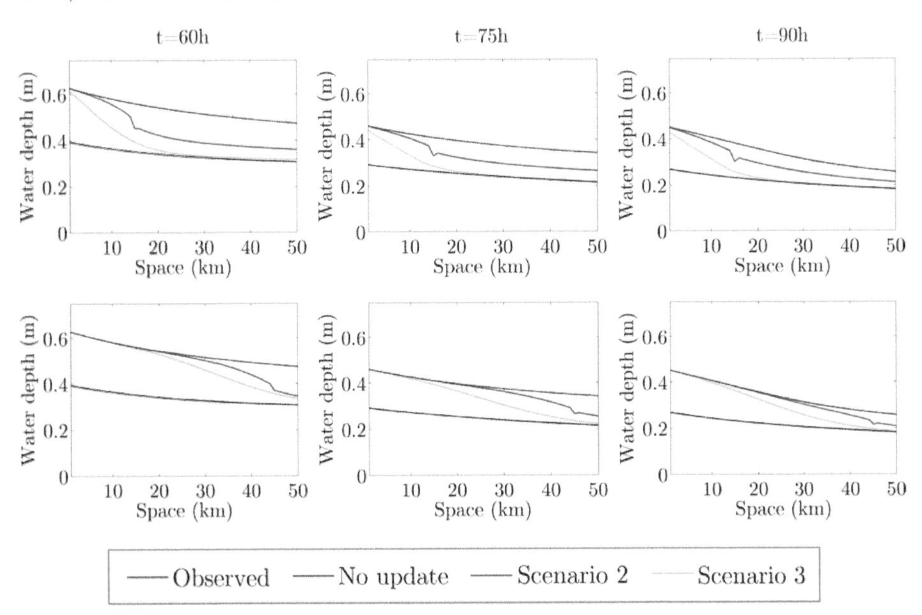

Figure 6.17. Water depth profile at three particular time step in case of assimilation point at 15km (first row) or 45 km (second row) during flood event A

However, such point is not the optimal sensor location. In fact, the minimum **P** value can change according to the location of the AP. One possible way to estimate the optimal sensor location might be to calculate the minimum value of P among the all possible sensor locations. However, diverse values of $\overline{\mathbf{M}_m}$ and $\mathbf{M}_b$ might affect the DA performances in different ways. For this reason Scenario 2 and 3 are introduced.

Figure 6.17 shows the water profile along the river at three different time steps during flood event A in case of Scenario 2 and 3 with AP at 15km (1 row) and 45km (second row). High error in the boundary conditions tends to better improve water profile when the AP is closer to the boundary conditions. This can be related to the fact that Muskingum-Cunge propagates the updated states only downstream and not upstream, so in case of AP at 45km the updated at the upstream sections can be mainly related to the KF and not to the additional propagation effect of the Muskingum-Cunge. In addition, in case of Scenario 3 it can be seen how the states update seems to be continuous in space while with Scenario 2 an abrupt change in the water profile is localized at the AP. This can be explained from Figure 6.18. In fact, in case of Scenario 2 (model error is higher than boundary error) the model update, expressed in terms of Kalman gain **K**, is localized at the AP with small improvement at the boundary conditions. On the other hand, in case of Scenario 3 (model error is smaller than boundary error) it can be pointed that the highest value of **K** (i.e. maximised gains by KF), in case of AP at 15km and not optimal sensor location, is achieved at the boundary location and then such gain is propagated downstream, generating the continuous update that is showed in Figure 6.17. Similar results are obtained when WD observations are assimilated at 30km and 45km. Such high values of **K** are not used to estimate the optimal sensor location. In fact, the main goal of this chapter is to show how, states, **K** and **P** changes according to model and boundary errors for assigned location of StPh sensors

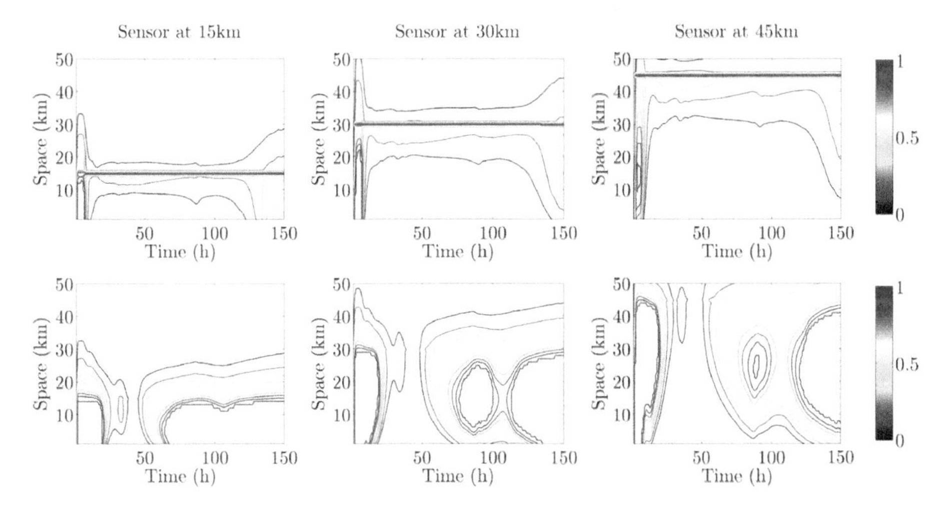

Figure 6.18. Representation of the Kalman gain in case of Scenario 2 (first row) and 3 (second row)

In Figure 6.19 the NSE values in each section of the river are estimated comparing observed (synthetic) and simulated WD in case of AP at 15km, 30km and 45km. Assimilation of WD at one specific location tend to rapidly increase NSE around that point in case of Scenario 2. In addition, downstream the AP, NSE increases up to an asymptotic value. Such value is higher in case of AP close to the reach outlet. This shows how, in case of model error higher then boundary error, it is recommended to locate the WD sensor close to the reach outlet.

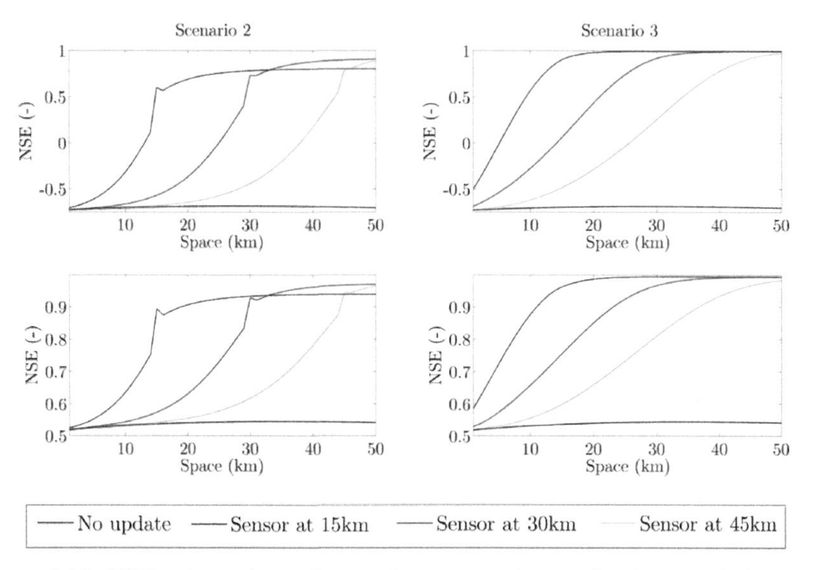

Figure 6.19. NSE values along the synthetic river during flood event A (first row) and B (second row) considering three assimilation point at 15 km, 30 km and 45 km

In fact, upstream boundary is considered accurate so no updated is performed by the KF next to it, as demonstrate in Figure 6.18. Opposite results are achieved with Scenario 3, where the NSE values obtained for the 3 AP tend to the same asymptotic value while close to the upstream boundary the NSE has high value for AP close to the boundary. This pointed out that, in case of model error lower than boundary error (Scenario 3), the sensor should be located closer to the upstream boundary in order to achieve a good model improvement along the whole river reach. Moreover, locating the sensor upstream will give additional response time in order to predict the WD at the outlet of the river reach, as demonstrated in the next section. Opposite results are obtained in case of Scenario 2.

Experiment 6.2.2: Bacchiglione River

Water depth observations from hypothetical StPh sensors, assumed at different locations, are assimilated into the Muskingum-Cunge model applied in each reach of the Bacchiglione River. The results reported in Figure 6.20 show the sensitivity of the model performances, at PA, to the assimilation of WD observations in different river reaches during different flood events. Overall, assimilation within the

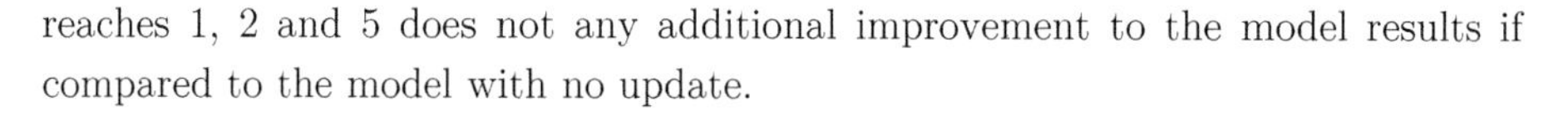

reaches 1, 2 and 5 does not any additional improvement to the model results if compared to the model with no update.

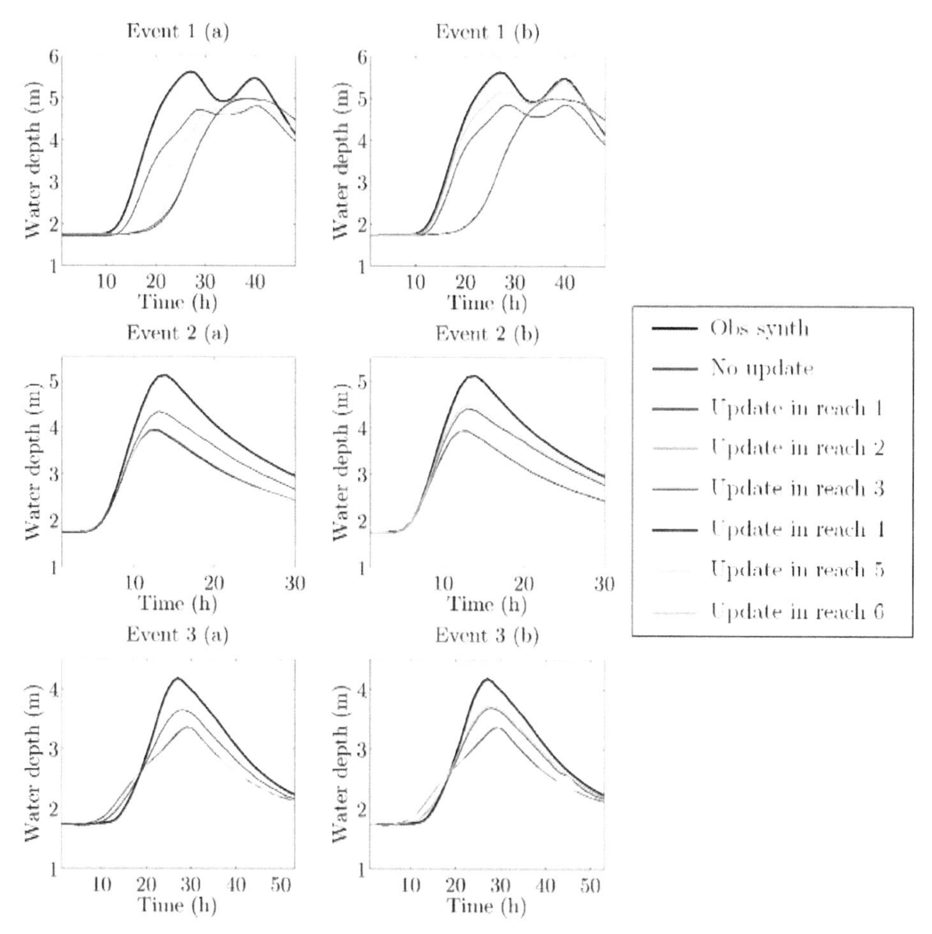

Figure 6.20. Simulated hydrographs at PA obtained assimilating WD observations from sensors at the first cross section in the different reaches of the Bacchiglione River.

On the other hand, as in case of the synthetic river, accurate 'water level' prediction are achieved assimilating WD observations by means of KF in reaches 3, 4 and 6. This can be related to the fact that these reaches are located upstream the reaches 3, 4 and 6 proving a lower contribution in the overall model improvement at PA. Among the remaining reaches, reach 6, located in the downstream part of the catchment, is the one which allow achieving the best model update. It is worth noting that the impacts of sensor positioning are evaluated without assimilating all observations at the same time. If all information is considered, impacts of some additional measurements might be marginal or redundant.

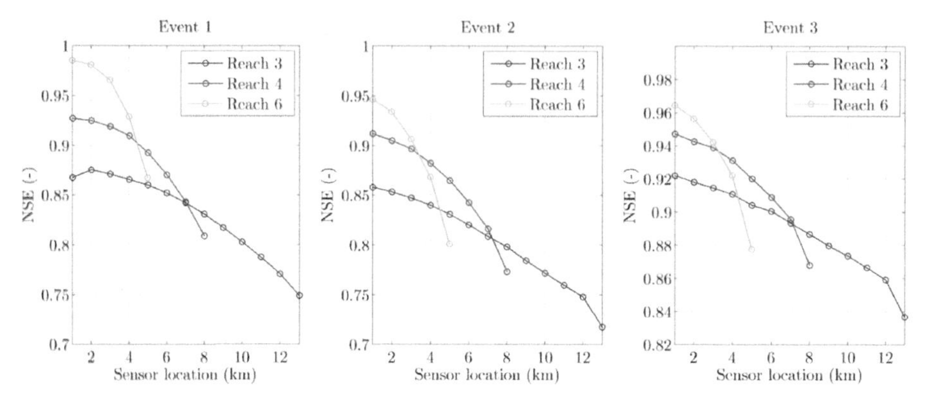

Figure 6.21. Representation of the NSE with respect to the different sensor location along the reach 3, 4 and 6 during three different flood events

Figure 6.21 shows the NSE obtained positioning the hypothetical sensors at different location along the reaches 3, 4 and 6. As previously demonstrated, reach 6 is the one that provides better model improvement in case of sensor located upstream. In fact, the higher values of NSE are obtained assimilating WD observations at the upstream of each river reach as a consequence to the fact that boundary conditions have higher error than model itself. Moreover, during flood event 1, moving the sensors towards the outlet of reach 6 induces a significant decreasing on the model performances up to 12% of the NSE, while in case of reach 3 and 4, moving the sensors 5km downstream as in case of reach 6, such worsening it is up to 1% and 4% respectively. However, the overall worsening in model performances from upstream to the outlet downstream for the 3 reaches in case of the 3 different flood event is reported in Table 6.3.

In case of different lead time values, model improvements are different according to the location in which the WD observation is assimilated. In Figure 6.22 the WD observations are assimilated until a certain time of forecast (TOF) while after that the model is running in prediction mode, i.e. without update, in case of flood event 1. The sharp decrease of the hydrograph after TOF may be due to underestimated forecast forcing.

The results of Figure 6.22 pointed out how reach 6 tends to lose the effect of WD assimilation faster than reaches 3 and 4. This is due to the travel time along reach 6, in fact, considering an average flow velocity of 1m/s and the 5km-length of the reach, the MC model will lose the effect of the assimilation procedure after 1.5h, as demonstrated in Figure 6.22. On the other hand, reach 3, even if it induces a lower model improvement than the one obtained assimilating WD within reach 6, is the one that provides the longest memory in the system due to an average travel time of about 5.5h. A compromise between reach 3 and 6 is achieved in case of river reach 4.

Table 6.3 Model performances from upstream to downstream for reaches 3, 4, and 6 during flood events 1, 2 and 3

		Event 1	**Event 2**	**Event 3**
Reach 3	Upstream	0.868	0.858	0.922
	Downstream	0.749	0.748	0.837
	Improvement (%)	**-0.158**	**-0.196**	**-0.102**
Reach 4	Upstream	0.927	0.912	0.947
	Downstream	0.809	0.773	0.868
	Improvement (%)	**-0.146**	**-0.179**	**-0.091**
Reach 6	Upstream	0.986	0.946	0.965
	Downstream	0.867	0.801	0.878
	Improvement (%)	**-0.136**	**-0.181**	**-0.099**

In Figure 6.23, a representation of the model performances expressed in terms of NSE are reported as a function of the lead time for the all flood events. The different lines indicate the locations within the single river reach in which the WD observations are assimilated.

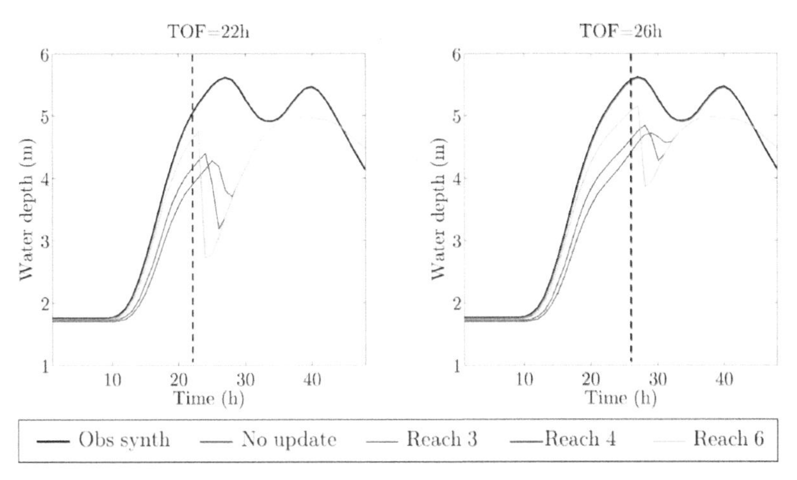

Figure 6.22. Water depth hydrographs at PA obtained assimilating WD observations in the first cross section of reaches 3, 4 and 6 considering two value for the Time Of Forecast (TOF) during flood event 1.

For example, the black line in reach 3 indicates the observation is assimilated at 2km, i.e. the second cross section of that reach since Δx is equal to 1000m. From this figure it can be observed a fast reduction in the model performances considering reach 6, in which the model tends to the NSE obtained with no update after 2 hours. Reaches 3 and 4 tend to the model performances without update with higher lead time than reach 6, in accordance to the considerations previously drawn. The choice of the optimal location of sensors should be reflected in an optimization between NSE and lead time. It is worth noting that the low NSE value achieved in case of MC without any update is due to the underestimation of the forecasted precipitation which induces lower values in the discharge and WD.

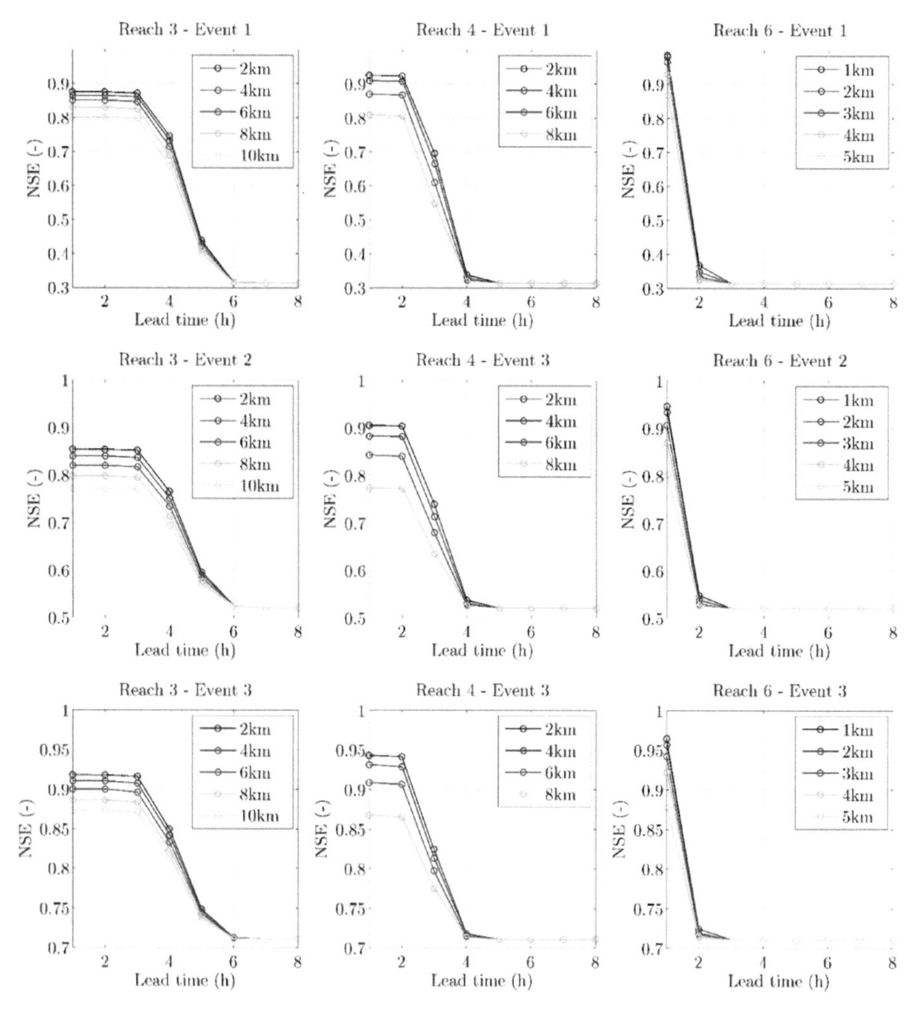

Figure 6.23. NSE values with respect to the lead time at PA considering different sensor locations (different colours in the graph) along reach 3, 4 and 6

Real world experiments

In order to validate the results of the real world experiments, the NSE, R and Bias statistical indexes are calculated at the Bacchiglione outlet (PA) assimilating WD observations at PM and PA respectively (Figure 6.24). It is worth noting that these analyses are related to flood event 1 which is the one with highest magnitude among the ones considered. In Figure 6.24 it is pointed out that overall assimilation of WD observations tend to underestimate the observed WD values while a good correlation it is found between observed and simulated WD. In particular, assimilation at PM provides the lower NSE and R with respect to the model updating using WD observations at PA. Assimilation at both PA and PM shows an improvement in the correlation R and slight increasing in the NSE values. However, Bias value is reducing (optimal value should be 1) if compared to the one obtained assimilating WD only at PA.

This can be related to the fact that simulated WD at PM, with no model updating, is higher than observed WD as showed in Figure 2.12. At the moment observed WD is assimilate at PM this induces a reduction in the simulated WD that, combined with a uncertain estimation of the internal boundary conditions I6 and I7 (see Figure 2.10) induces an underestimation at PA as reported in the hydrographs and statistical indexes of Figure 6.24. In Figure 6.25 the Taylor diagram, representing the standard deviation, correlation coefficient R and root mean squared difference of simulated and updated hydrographs, summarise the results previously described.

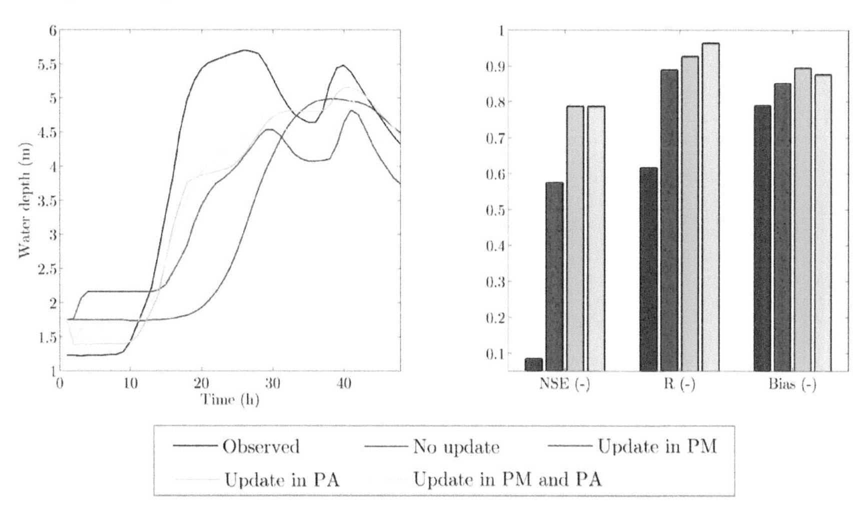

Figure 6.24. Water depth hydrograph and statistical indexes at PA obtained in the real world experiment along the Bacchiglione River

Figure 6.26 highlights the benefits of assimilating WD observations upstream the target point (PA). In fact, while assimilation at PA is useful to improve NSE, R and Bias for low values of lead time, the assimilation at PM helps to increase the memory of the system and improve model performances for high values of lead time up to 4 hours, i.e. the travel time from PM to PA.

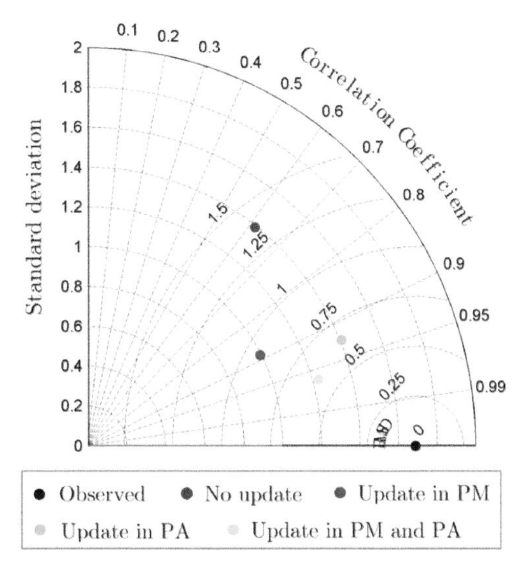

Figure 6.25. Taylor diagram representing the statistics of the hydrographs obtained in the real world experiment along the Bacchiglione River

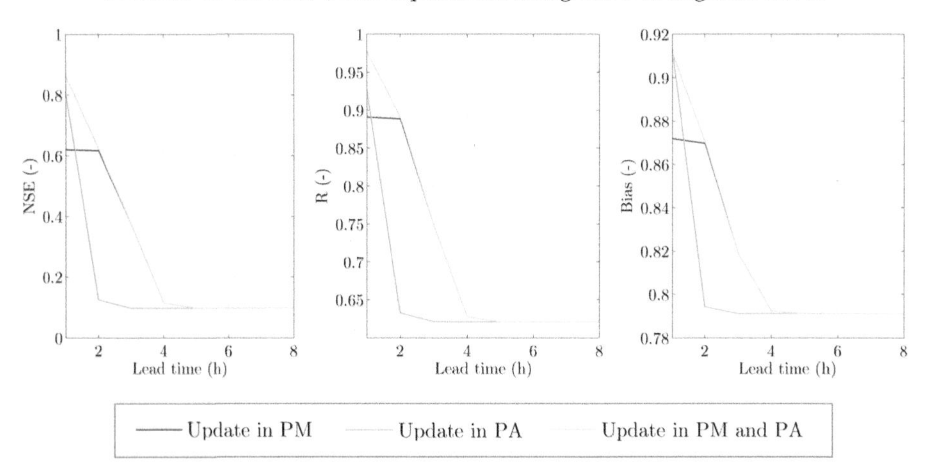

Figure 6.26. NSE values with respect to the lead time at PA during the real-world experiments along the Bacchiglione River

6.5 CONCLUSIONS

This chapter evaluated the effect of a) different DA approaches for assimilating streamflow observations from existing StPh sensors and b) assimilation of water depth observations coming from spatially distributed of StPh sensors. For this reason two Muskingum-Cunge models, having 2 and 3 parameters, were implemented along the Trinity and Sabine rivers (Experiment 6.1) and a synthetic river and the Bacchiglione River (Experiment 6.2). In case of Experiment 6.1, different DA methods are used, while in Experiment 6.2, because of the linear nature of the MC model, a KF is applied. In Experiment 6.1, real streamflow observations are assimilated within the 3-parameter Muskingum model, while in Experiment 6.2 synthetic water depth observations are used due to the impossibility of having distributed observations along both river reaches.

Experiment 6.1 showed that, overall, assimilation of streamflow observations improves model performance. In particular, in the case of the lumped model (Experiment 6.1.1), DI and AEnKF provide the best model improvement among all the used DA methods. The data assimilation methods, and in particular Kalman filtering approaches, are noticeably sensitive to the definition of model error. Also in the case of the distributed 3-parameter Muskingum model (Experiment 6.1.2), the highest model performances are achieved using AEnKF. Increasing the number of past observations used in AEnKF increases the model performance. On the other hand, DI is not as effective as in the case of the lumped model structure. This is due to the fact that, using this method, model states are updated only at the assimilation location, while with the Kalman filtering approach the update is performed along the whole river reach because of the distributed nature of the Kalman gain and covariance matrix. In addition, because of the model structure, the updating effect is more significant in the downstream reaches than in upstream reaches for all DA methods used in this chapter.

Experiment 6.2.1 showed that diverse values of model error might affect the assimilation performances in different ways. In particular, for high error in the boundary condition, the water profile tends to be better predicted when the assimilation point is closer to it and a smooth update is achieved downstream of the assimilation point. On the other hand, in the case of model error higher than boundary error, an abrupt update is obtained at the water depth sensor location and good model performances are achieved if the StPh sensor is located close at the reach outlet. That is why, in this last case, it might be suggested to locate the sensor downstream of the river reach to maximise the model improvement at the river outlet.

Experiment 6.2.2 showed a good fit between the flows and water depths estimated at two different locations (PM and PA) along the Bacchiglione river using MIKE11 and the Muskingum-Cunge model in case of forecasted and measured boundary conditions during flood event 1. Values of NSE, R and Bias of around 0.95, 0.995 and 1.05 are obtained comparing the water depths estimated with the two models at PA, Vicenza, in case of forecasted boundary conditions. The results obtained in the Bacchiglione River showed that only the assimilation within the reaches 3, 4 and 6 provides additional improvement to the model results at the outlet (PA), as compared to the model with no update. Due to the fact that upstream boundary conditions have higher error than the routing model itself, high values of NSE are obtained assimilating water depth observations close to the upstream end of each river reach, as previously demonstrated. Among the previous three main reaches, reach 6 provides the best model performances. In case of forecasting, reach 6 tends to lose the assimilation effect faster than reaches 3 and 4 due to its shorter travel time. In fact, StPh sensors located at the upstream part of these reaches ensure additional lead time, up to 6 hours, for the prediction of water depth at the reach outlet. For this reason, the choice of the optimal location of StPh sensors should be a compromise between best NSE value and prediction capability of the model itself.

7

Assimilation of synchronous data in a cascade of models

Chapter 4 and 5 were devoted to hydrological models and Chapter 6 to hydraulic models. However, EWS often work in a cascade of models containing both hydrological and hydraulic models. This chapter illustrates the benefits of assimilating CS observations from a heterogeneous network of StPh, StSc and DySc sensors in a cascade of hydrological and hydraulic models in the Bacchiglione catchment during one particular flood event. A standard Kalman filter is implemented within the lumped conceptual hydrological model and along the Muskingum-Cunge model, described in chapter 2, to assimilate CS streamflow and water depth observations respectively. CS data are represented by realistic synthetic model-based observations having random accuracy and spatio-temporal coverage. Different experiments are performed in order to assess the effect of different sensor types on the model prediction at PA (Vicenza) during flood event of May 2013. In particular, realistic scenarios of citizen engagement are introduced in order to properly assess the effects of CS observations on the DA method and model results.

This chapter is based on the following peer-reviewed journal publication:

Mazzoleni M., Cortes Arevalo V.J., When U., Alfonso L. and Solomatine D.P. (2015) Assimilation of crowdsourced observations into a cascade of hydrological and hydraulic models: The flood event of May 2013 in the Bacchiglione basin, Hydrology and Earth System Sciences, In preparation for Hydrology and Earth System Science

7.1 INTRODUCTION

In the last decades, different attends have been made in order to improve flood model predictions by mean of model updating techniques. However, most of the previous studies have focus on the assimilation of a single hydrological variable per time, while few studies have examined methods of assimilating multiple observations from different sensors and of different hydrologic variables (Montzka et al., 2012; Andreadis et al., 2015).

One of the first attempts to assimilate observations coming from multiple sources is proposed by Aubert et al. (2003). In that study, daily soil moisture and streamflow data from StPh sensors are used to update the states (soil and routing reservoirs) of a conceptual lumped rainfall-runoff model by means of an extended Kalman Filter. The authors showed that, overall, streamflow prediction is improved by the coupled assimilation of both soil moisture and streamflow individually. In McCabe et al. (2008), multiple remote sensing-based observations as soil moisture (from AMSR-E), precipitation rates (from TRMM) and surface heat fluxes (from MODIS), are integrated with a modelling system in order to predict water and energy cycle. The authors pointed out that the use of multi-sensor observations can be used to achieve a reliable description of both the land surface water cycle. In a similar study, Pan et al. (2008) assimilated, using EnKF and particle filter, satellite remote sensing data from multiple sensors within a variable infiltration capacity model, a land surface microwave emission model and a surface energy balance system model in order to achieve a good estimates of the water budget at the regional scale. The satellite products include the Tropical Rainfall Measurement Mission (from TRMM), TRMM Microwave Imager (from TMI), and Moderate Resolution Imaging Spectroradiometer (from MODIS). The authors demonstrated the feasibility of assimilating multi-sources remote sensing data and a good model improvement over the predictions without any model update. Lee et al. (2011) assimilated streamflow and in situ soil moisture observations from StPh sensors in a distributed hydrological model. They carried out both synthetic, with no structural and parametric uncertainties in the hydrological models, perfectly known precipitation and potential evaporation, and real-world experiment to assess the DA performances. In case of synthetic experiments, under the introduced assumptions, good model improvements are achieved assimilating both streamflow and soil moisture. However, in the real-world experiment, assimilating streamflow and in situ soil moisture data provided little improvement. This might be due to the combination of structural and parametric errors in the hydrologic models and large observational error in case of real-world experiment. In Pipunic et al. (2013), multiple data types derived from a range of remotely sensed products are merged with a land surface models demonstrating the value of multi-observation assimilation for improving heat flux prediction. Andreadis et al. (2015) developed a multi-sensor and multivariate data assimilation forecast system RHEAS (Regional Hydrologic Extremes Assessment System), based on the variable infiltration capacity model, with the initial focus on forecasting drought characteristics. Different satellite observations, such as soil moisture (from AMSR-E), water storage (from GRACE), evapotranspiration (from MODIS), and precipitation (from TRMM), are assimilated in order to update the initial model states. In-situ observations are used as benchmark. Lopez Lopez et al. (2015), studied the effect of assimilating both streamflow, from 23 StPh gauging stations, and satellite soil moisture, from AMSR-E, observations in a global hydrological model in case of either coarse- or high-resolution meteorological observations by mean of an EnKF. Overall, the joint assimilation of both soil moisture and streamflow leads to a 20% reduction of the RMSE. In particular, the single assimilation of soil moisture observations induces a larger model improvement. In Rasmussen et al. (2015), groundwater head and stream discharge values are assimilated by means of an ensemble transform Kalman filter within an integrated hydrological model. In particular, such study focuses on the relation between ensemble size and filter performances. The authors demonstrated that in case of few assimilated groundwater head observations a larger ensemble size is needed. In addition, a larger ensemble size than the one used to assimilate groundwater head observations is need in case of assimilating discharge observations.

The recent technological development has stimulated the use of low-cost StSc and DySc sensors used by citizen to derived hydrological CS observations, such as precipitation and water level. However, CS observations are not integrated, in a real-time fashion, within hydrological and/or hydraulic model in order to improve flood forecasting. The reason behind this decision are related to the random accuracy of CS observations and their random spatio-temporal coverage. Due to the complex nature of the hydrological processes, spatially and temporally distributed CS data can be used to ensure a proper flood prediction. However, no previous study addressed the issue of assimilating CS from heterogeneous sensors in flood forecasting applications. Recently, Schneider et al. (2015) reported an example of data fusion used to provide a combined concentration field by regressing dynamic air quality observations against model data and spatially interpolating the residuals.

For this reason, the main objective of this chapter is to assess the benefit in assimilating CS observations derived from a distributed network of StPh, StSc and DySc sensors within a cascade of hydrological and hydraulic models during one particular flood event. In particular, different experiments are carried out in order to assess the single effect of StPh (Experiment 7.1), StSc (Experiment 7.2) and DySc (Experiment 7.3) sensors in the cascade of hydrological and hydraulic models. In the last experiment, Experiment 7.4, assimilation of CS observations from all sensors is carried out, considering a realistic assumption of citizen engagement in providing such uncertain data.

7.2 METHODOLOGY

In this chapter, synchronous CS water depth observations, coming from a heterogeneous network of StPh, StSc and DySc sensors, are assimilated within the hydrological and hydraulic models implemented in the Bacchiglione catchment (see section 2.3.2) during the flood event of May 2013. In order to assimilate such synchronous observations a standard KF is used. As in case of the previous chapters, realistic water depth observations are used due to the fact that CS observations ae not available at the time of this chapter. Below, a short description of the DA method, the definition of the observation and model error and the estimation of the realistic observations are reported due to the fact that most of them are already introduced in previous chapters.

7.2.1 Data assimilation method

A standard version of the KF is implemented in both hydrological and hydraulic model of the Bacchiglione catchment. In particular, in case of hydrological model, the Eqs.(2.6) of x_{sur} and the ones from Eq.(3.6) to Eq.(3.10) are used to assimilate synthetic synchronous streamflow observations within the model itself. However, due to the fact that the observations vector in the Muskingum-Cunge model is expressed in terms of river flow, the realistic values of WD are converted in streamflow using the manning equation for the natural river cross-section available to then be assimilated within the hydraulic model. The KF implemented within the Muskingum-Cunge model is based on Eq.(2.22) and from Eq.(3.6) to Eq.(3.10).

7.2.2 Observation and model error

The estimation of the observation error is performed in the same way previously described. In fact, the value of the error R_t is calculated based on Eq.(4.4), where assumes different value according to the type of considered sensor, as is showed below.

As described in the previous chapters, the correct definition of model covariance matrix **S** can affect the DA performances and the consequent correctness of the flood predictions. For the hydrological model, such matrix is estimated as the diagonal matrix with constant value equal to 10^2 as described in chapter 5.

On the other hand, in case of hydraulic model not only the model error itself should be calculated but also the one related to the boundary conditions. For this reason, the same procedure described in section 6.2.2 (page 134) is followed. In this way, the error of the Muskingum-Cunge model is set equal to 39.7 m^6/s^2 in each river reach, comparing the simulated streamflow with Muskingum-Cunge and MIKE11 model. Errors at the upstream boundary conditions, $\mathbf{M_{b,1}}$, $\mathbf{M_{b,2}}$ and $\mathbf{M_{b,5}}$ are calculated as standard deviation between observed (synthetic) and simulated streamflow hydrographs. In this way, the effect of assimilation of streamflow observations in hydrological models is accounted. In fact, lower boundary error is expected because of the better model performances with KF. The errors at the boundary conditions of reaches 3, 4 and 6, i.e. $\mathbf{M_{b,3}}$, $\mathbf{M_{b,4}}$ and $\mathbf{M_{b,6}}$ are estimated following the approach described in section 6.2.2.

7.2.3 Generation of synthetic observations

As previously describe in Chapter 5 (section 5.2.3, page 112), realistic CS observations are assimilated instead of real-time data. Such synthetic observations are based on model results considering only one flood event occurred in May 2013. Measured time series of precipitation is used to estimate the synthetic observed hydrographs at the outlet of each sub-catchment of the Bacchiglione catchment. Such streamflow values are used both as synthetic observations for the StPh and StSc sensors and as input in the hydraulic model of the Bacchiglione River. The output of the Muskingum-Cunge model are used as realistic synthetic water depth observations for each x (1000m) of the river reach. On the other hand, forecasted precipitation value are used as input in the hydrological model to estimate the simulated water depth at the outlet of the sub-catchment, along the river reach and, in particular, at the target point of PA (Vicenza). It is worth noting that, in this case, hydrological and hydraulic model time step is set equal to $0.9 \cdot x$, i.e. 900 seconds. This means that realistic CS observations are provided synchronously assimilated every 900 second within the models.

7.3 Experimental setup

7.3.1 Experiment 7.1: Assimilation of data from static physical (StPh) sensors

In this experiments, WD observations coming from StPh sensors are assimilated within the hydrological (sensor StPh1) and hydraulic (sensor StPh2) and (sensor StPh3) models of the Bacchiglione catchment. The observations are assumed to be synchronous and regular in time. In particular, because of the high accuracy of such sensors, the coefficient of Eq.(4.4) is considered equal to 0.1, constant in time and space. The assimilation of WD observations is firstly performed considering a single sensor per time and then all the StPh sensors at the same time. Model performances, expressed in terms of NSE, are calculated for different lead time values, up to 24 hours.

7.3.2 Experiment 7.2: Assimilation of data from static social (StSc) sensors

In Experiment 7.2, only assimilation of WD observations from StSc sensors is considered. Beside for StSc1, 2 and 6, located in sub-catchment A, B and C respectively, the other sensors are located along the river reaches of the Bacchiglione catchment. As for the WD observation from StPh sensors, also the ones form StSc sensors are considered synchronous in time. However, on contrast to the observations from StPh, the ones from StSc are not regular in time since are strictly related to the citizen behaviour.

For this reason, different values of citizen engagement (called Eng in the following) are considered. Such engagement, closely related to the intermit nature of the WD observations, can be considered as the probability to receive an observation at a given model time step. This means that in case of Eng=0.4 there is 40% of probability to have an observations or not at a given model time step. In fact, in case of Eng=0 no observation is assimilated and the semi-distributed model is run without any update. On the other hand, for Eng=1, observations are available every time step and such situation is analogous to the case of observation from StPh sensors regular in time. In Chapters 4 and 5, intermittent streamflow observations with Eng=0.5 are considered.

Observation error is defined as in previous chapters using Eq.(4.4). In particular, the coefficient is considered equal to be a random stochastic variable uniformly distributed in time t as $U(_{min}, _{max})$, where α_{min} and α_{max} are set to 0.1 and 0.3 respectively to account for the low and high observational noise. As in case of

Chapter 5, the value of α for each StSc sensor is considered only function of the time t since the location of the sensor is assigned and fixed. Assimilation of WD observations in case of different combination of sensor availability in the different sub-catchments and river reaches is performed.

7.3.3 Experiment 7.3: Assimilation of data from dynamic social (DySc) sensors

In the experiment 7.3, the assimilation of WD observations coming from DySc sensors (described in Chapter 1, page 17) is considered. In this case the CS observations can be sent without the use of the static reference tool as in case of StSc sensors but only with the dynamic device (e.g. smart phone). The main 2 differences between StSc and DySc sensors are that: 1) DySc sensor locations vary every time step along the river reaches in contrast to StSc sensor which locations are considered constant in time. In fact, in case of DySc sensors, the mobile sensor might provide observations in different random places due to the fact that there is no need of a static reference tool to measure the water depth; 2) uncertainty in the observations provided by DySc sensors is higher than the ones from StSc sensors. This is due to the fact that, for a non-expert, it might be difficult to estimate water depth in a river without any reference device as in case of StSc sensors. For this reason, citizen might provide observations of distance between water profile and river bank. Such information is then used by the modeller to calculate water depth knowing the distance from river bank and thalweg (from available natural section). This procedure introduces high uncertainty in the estimation of water depth.

However, in this chapter, due to the lack of real-time observations from dynamic sensors, a synthetic WD value is considered instead of the distance between profile and river bank. Such realistic WD observations are then assimilated only in the hydraulic model of the Bacchiglione River. This is due to the fact that WD observations are easier to be integrated within hydraulic model than hydrological. In fact, water depth observations should be converted into streamflow values, for example by means of a rating curve, in order to be assimilated within the hydrological model. However, due to the distributed and dynamic nature of the DySc sensors, it would be very difficult to assess the rating curve in a random position along the river reach. Also in this experiments, different values of engagement, as previously described in Experiment 7.2, are accounted. In order to represent the uncertain nature of WD observations from DySc sensors, the coefficient α of Eq.(4.4) is considered to be a random stochastic variable, uniformly distributed in time t and space as $U(\alpha_{min},\alpha_{max})$, where α_{min} and α_{max} are set to 0.2 and 0.5. In addition to the random error of the WD observations from DySc sensors, a systematic error is also accounted by means of different values of observations bias estimated as:

$$WD_t^{biased} = WD_t^{synth} + \gamma_t = WD_t^{synth} + QWD_t^{synth} \cdot U(\gamma_{min}, \gamma_{max}) \tag{7.1}$$

where γ is a random stochastic variable function of the time, having minimum and maximum values γ_{min} and γ_{max} are equal to:

Table 7.1 Minimum and maximum values γ_{min} and γ_{max} in case of 4 different cases of observation bias

	γ_{min}	γ_{max}
Bias 1 (γ_1)	0	0
Bias 2 (γ_2)	-0.3	0.3
Bias 3 (γ_3)	-0.3	0
Bias 4 (γ_4)	0	0.3

where scenario 1 represent the case of no bias, scenario 3 of underestimation and scenario 4 of overestimation of the real WD value.

7.3.4 Experiment 7.4: Realistic scenarios of engagements

In this experiment, all the StPh, StSc and DySc sensors are considered. However, the engagement level is estimated in a more realistic and complex way. In fact, in the previous experiments, engagement is considered as random values varying from 0 to 1. However, in Experiment 7.4, such engagement level is considered a function of the population distribution within the Bacchiglione catchment. In order to estimate such engagement the following 3-steps procedure is proposed.

Step1: Estimate of the citizen active area. A 500m buffer around each sub-river reach of 1000m (Δx of the Muskingum-Cunge model) is used to identify the area in which the active population which might provide CS observations using DySc sensors (see Figure 7.1). It is in fact assumed that citizens located far more than 500m from the river are not contributing to the collection of CS observations. In case of StSc sensor, the active area is assumed as a circle of 500m radius with centre in the sensor itself. Land cover map are used to identify the main urban area from which citizens might provide CS observations of WD within the buffer previously estimated (see Figure 7.1).

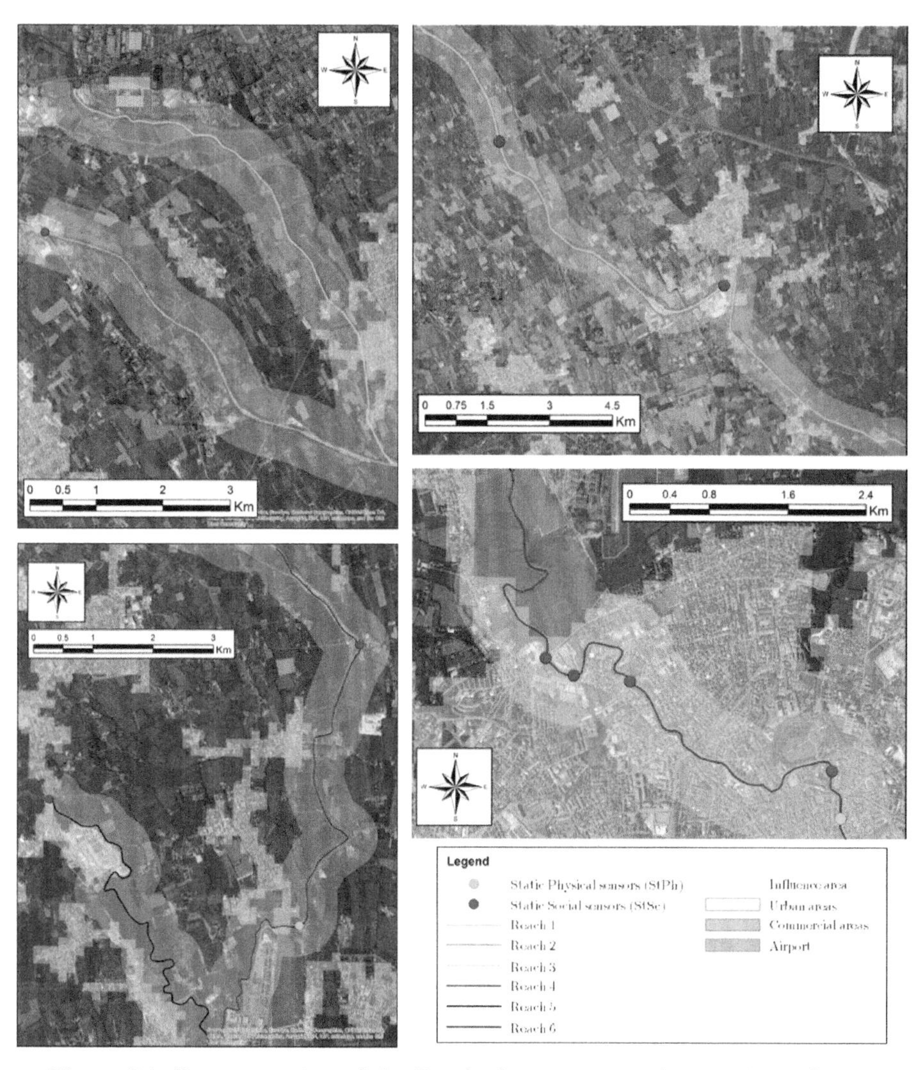

Figure 7.1. Representation of the Bacchiglione river reaches, land use (Corine Land Cover, 2006), location of the StSc and StSc sensors and the 500m buffer

Step 2: Estimate of the number of active citizens. The population density for the different municipalities along the different river reaches is used to estimate the number of citizens within the 500m buffer for each sub-river reach of 1000m in which urban area are located. In case of no urban area, an engagement value equal to 0 is considered. It is worth noting that not all the citizens would be able to provide CS observations due to the fact that only part of them is using mobile phone. According to Statistica (2016), the mobile phone penetration in Italy in 2013, the year of the flood event analysed in this chapter, is about 41%. For this reason, in order to estimate the active population, the number of citizens enclosed between the 500m buffer and 1000m of river sub-reaches, previously estimated, is multiply with such percentage. The results of these analysis are reported in Table 7.2 in case of StSc sensors and

Table 7.3 for DySc sensors. In case of main urban areas contained in more than one sub-reach, as in case of reach 6 (km 3-4-5) in

Table 7.3, the active citizens are divided by the number of sub-reaches (3 in the previous case of reach 6).

Table 7.2 Estimate of the active population which can provide CS observations of WD from StSc sensors

Sensor	Municipality	Active area (m^2)	Density ($inhab/km^2$)	Population (inhab)	Active citizens (inhab)
StSc–1	Schio	206828.3	597	124	51
StSc–2		71292.5		43	18
StSc–3	Malo	100733.8	491.39	50	21
StSc–4	Villaverla	359743.8	400	144	59
StSc–5	Caldogno	67310.9	720	49	20
StSc–6	Costabissara	421777.7	562.53	238	98
StSc–7	Vicenza	86543.9	319.49	28	11
StSc–8		241.450.9		77	32
StSc–9		415513.4		133	55
StSc–10		500000.0		160	66

Table 7.3 Estimate of the active population which can provide CS observations of WD from DySc sensors

Reach	Municipality	Active area (m^2)	Density ($inhab/km^2$)	Population (inhab)	Active citizens (inhab)
1(km6-7-8)	Marano Vicentino	608985.2	800	487	200
2 (km2)	Schio	39536.4	597	24	10
3(km8)	Villaverla	359743.8	400	144	59
3(km11)	Caldogno	232474.1	720	167	69
4(km2)	Dueville	30692.3	700.85	22	9
4(km3)	Caldogno	191987.6	720	138	57

4(km5)	Costabissara	292519.8	562.53	211	86
5(km1)		351920.7		198	81
5(km2)		119897.9		67	28
5(3-4-5)		212452.9		68	28
6(km1-2)	Vicenza	129815.9	319.49	41	17
6(km3-4-5)		1156964.3		370	152

Step 3: Estimate of citizen engagement curve. Knowing the hypothetical number of active citizen, it is now necessary to estimate their level of engagement. For this reason, six different scenarios of Maximum citizen Engagement Level (*MEL*), function of three diverse citizen behaviours (Gharesifard and Wehn, 2016) and the number of active citizens, are proposed.

In the first citizen behaviour, it is considered that citizens collect data for their own personal purposes. In this case, the MEL is low for low number of citizens while it grows following a logistic function, Eq.(7.2), for increasing number of people.

$$MEL = \frac{K \cdot P_o \cdot e^{r \cdot pop}}{K + P_o \cdot (e^{r \cdot pop} - 1)} + w \tag{7.2}$$

where *pop* is the population number, *r* is the growth rate, *K* is the carrying capacity assumed equal to 1 and *w* is a coefficient related to the third citizen behaviour which is explained below. In this chapter, P_o and *K* are set equal to 0.01 and 1 respectively. Two different values of *r* are considered in the next scenarios of maximum engagement.

In the second behaviour, citizens might decide to collect and share CS observations driven by a feeling of belonging to a community of friends with sharing interests and vision. In such case, it is assumed that maximum value of *MEL* is achieved for small population values while for increasing population such value is reducing following an inverse logistic function.

In the third and last citizen behaviour, weather enthusiast individuals, weather networks and related hobby-club might provide additional information to the one already accounted in the first and second behaviours previously described. The added value of such information is accounted in Eq.(7.2) by means of a coefficient *w*. In Table 7.4, the characterization of the different engagement scenarios, based on different values of the coefficient *r* and *w* related to the citizen behaviours, is reported. A graphical representation of these scenarios is reported in Figure 7.2.

Table 7.4 Engagement scenarios based on different citizen behaviours

Engagement scenario	Citizen behaviour	r	w
1	2	0.12	0
2	1	0.04	0
3	1	0.08	0
4	1+3	0.04	0.05
5	1+3	0.08	0.05
6	1+3	0.04	0.15

In the next analysis, different model runs (100) are performed considering random values of citizen engagement from 0 to the maximum value of engagement (MEL) according to the given engagement scenario and population. For example, considering scenario 4 and 60 inhabitants enclosed in a given river sub-reach, different model runs are performed for engagement values varying from 0 to 0.6. In case different CS observations coming at the same time from different sensors, only the most accurate observation, i.e. having the lower value of observation error R, is assimilated in the hydrological and/or hydraulic model.

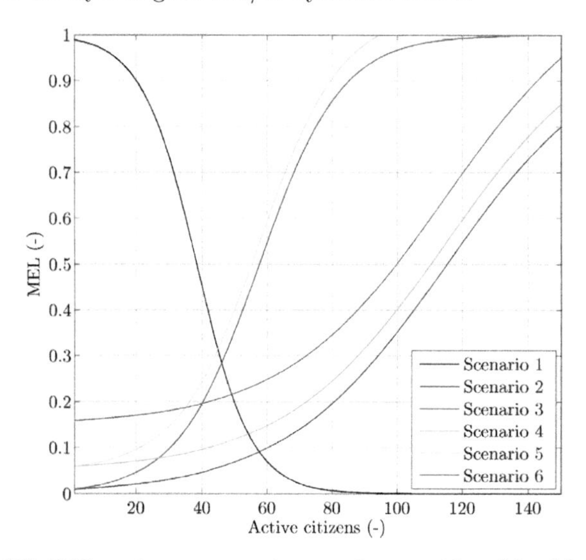

Figure 7.2. Different engagement scenarios considered in this chapter

7.4 RESULTS AND DISCUSSION

7.4.1 Experiment 7.1

Experiment 7.1 deals with the assimilation of streamflow and water depth observations from StPh sensors located in the hydrological (StPh1) and hydraulic (StPh2 and StPh3) models of the Bacchiglione catchment. As it can be seen from Figure 7.3, assimilation in hydrological model (StPh1) provides the best model improvement, in terms of water depth hydrograph at PA (Vicenza), if compared to the other StPh sensors. In particular, both flood peaks are well represented with assimilation from StPh1 sensor, while, with the other two StPh sensors, only the second simulated peak fits the observed values. Assimilation of water depth observations from StPh2 gives lower improvement than the assimilation from the StPh3 sensor, located close to the PA station. However, assimilation from StPh2 insures a better model prediction, expressed as NSE values, in case of high lead time value. This is due to the location of the StPh2 sensor, upstream StPh3, and the consequent high travel time (around 6 hours) required to reach the target point of PA. As it can be seen from Figure 7.3, travel time from StPh3 and PA is around 2 hours, after that, NSE drops to the value achieved in case of no model update.

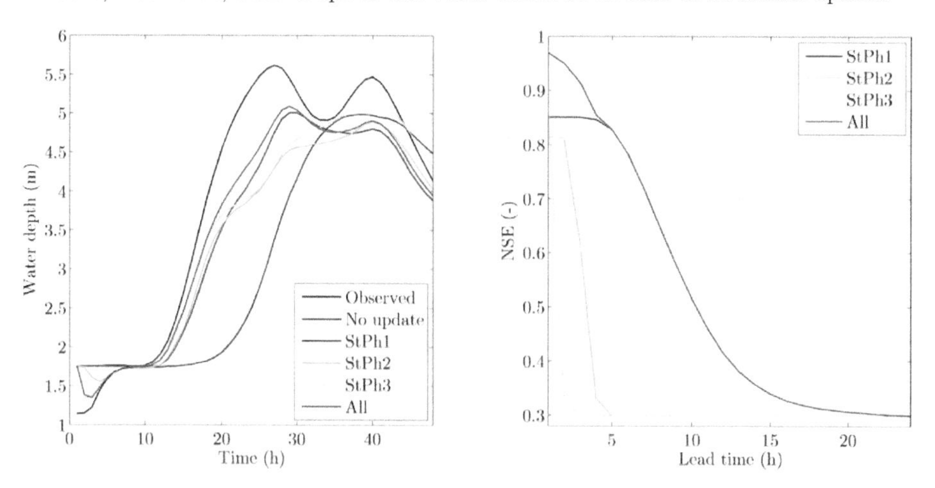

Figure 7.3. Assimilation of streamflow and water depth observations from StPh sensors in hydrological (StPh1) and hydraulic models (StPh2 and StPh3)

The compromise between model improvement and lead time value in case of different sensor location within the hydraulic model is showed in Chapter 6 (Experiment 6.2). Assimilation in hydrological model provides best model improvement also in case of high lead value. As expected, good fit of the simulated hydrographs and high NSE values are achieved from the assimilation from the distributed StPh sensors. In particular, up to 6 hours lead time NSE values are affected by the assimilation of streamflow and water depth observations from all StPh sensors, while after that, only StPh1 influences the model performance.

7.4.2 Experiment 7.2

In Experiment 7.2, only the assimilation of CS observations from StSc sensors is considered. Due to the fact that CS observations are not regular in time and they have variable accuracy, five engagement levels and random uniform values of the coefficient are considered. Several model runs (100) are performed to account for such random behaviour of CS observations. In each run, a specific value and arrival moment for each observation is considered and for such run a NSE value is estimated. From the sampling of such 100 NSE values, the corresponding mean (NSE) is calculated and showed in Figure 7.4 in case of assimilation from StSc sensors located at the outlet of the sub-catchment (hydrological model) or main river reaches (hydraulic models) where the sensors are located.

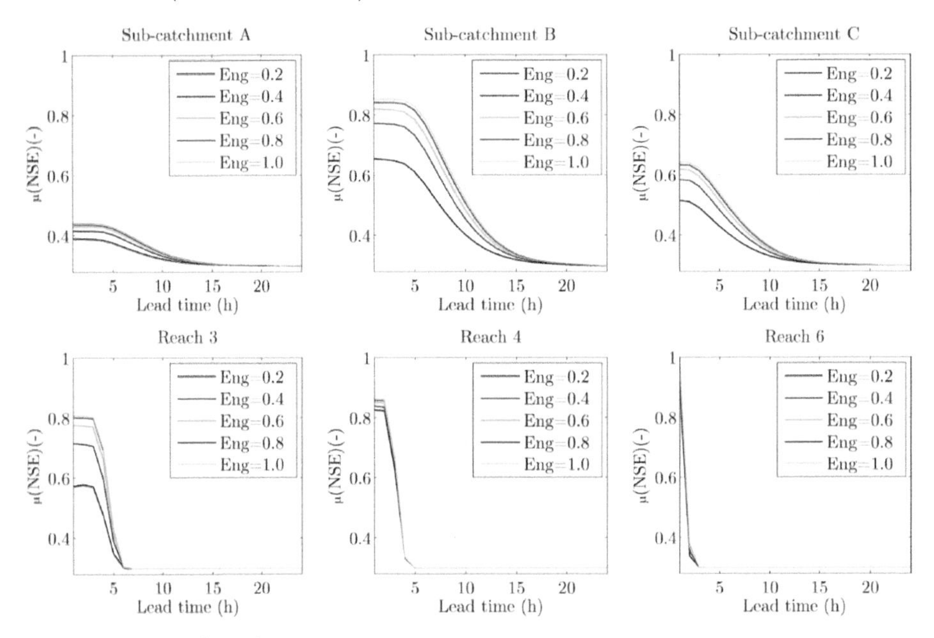

Figure 7.4. μ(NSE) obtained assimilating CS observations from different sub-catchments (first row) and river reaches (second row) in case of different engagement values

As in case of Experiment 7.1, different lead time values up to 24 hours are considered. From the results represented in Figure 7.4 it can be pointed out that assimilation from hydrological model allows to achieve good model predictions in case of high lead values. On the other hand, for short lead time, assimilation from StSc located in the river reaches (hydraulic model) induced high NSE values if the sensors are located close to the PA station. However, such improvement is not guaranteed for high lead time values due to the short travel time as showed in the previous sections.

As expected, for increasing engagement values NSE tends to increase as well. In case of engagement equal to 1, CS observations are received continuously every time step, while for engagement equal to 0.6 the CS observations have a 60% random probability to be received and then assimilated into the hydrological and/or hydraulic models. In Figure 7.5 the μ(NSE) values obtained assimilating CS observations derived from a combination of StSc sensors located in different sub-catchments and river reaches are represented for a lead time of 1 hour.

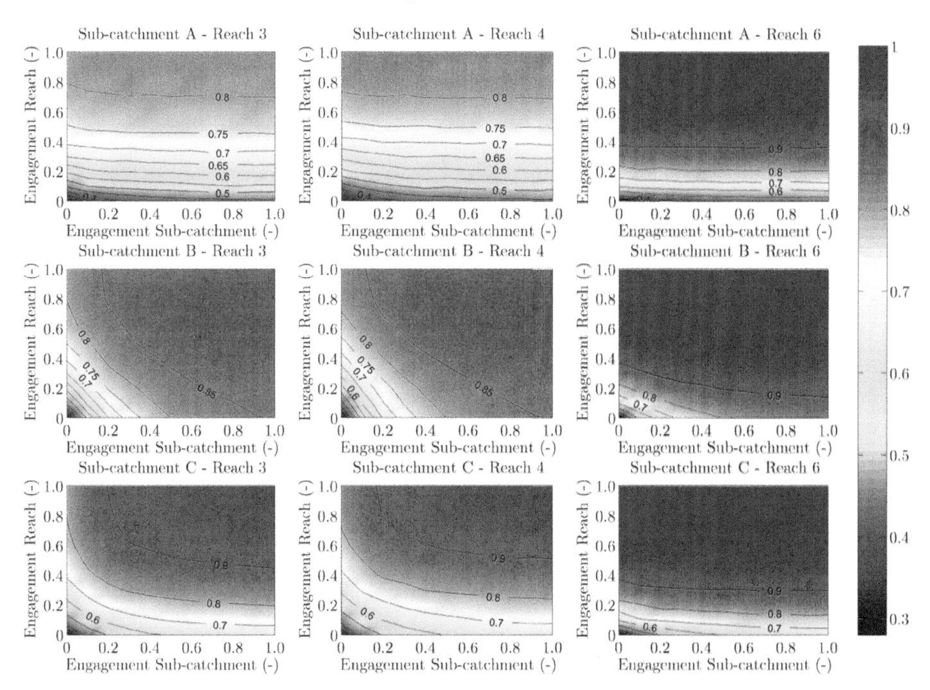

Figure 7.5. μ(NSE) values obtained assimilating CS observations from a combination of StSc sensors located in different sub-catchments and river reaches with 1-hour lead time.

For example, in the contour map located in the first row and first column, the NSE values obtained assimilating CS observations from sub-catchments A and river reach 3 are showed for different engagement values. Figure 7.5 pointed out that

NSE values are less affected by the assimilation of CS observations located in the sub-catchment A. In fact, from the first row of Figure 7.5 it is demonstrated that NSE values change only for different engagement values of StSc sensors along reach 3, 4 and 6, while constant NSE values are achieved for varying engagement values of the StSc2 (sub-catchment A). As previously showed, for low lead time value, NSE is higher in case of StSc sensors located in reach 6 rather than in the other river reaches 3 and 4. In case of assimilation in sub-catchment B, second row of Figure 7.5, higher NSE values are achieved if compared to the ones of the sub-catchment A (first row of the same figure). In particular, NSE values are mainly influenced by different engagement levels CS observation from sub-catchment B than river reaches 3. However, moving from upstream (reach 3) to downstream (reach 6) it can be observed a switch in the model behaviour, with an increasing influence of engagement in StSc sensors located close in river reach close to the PA station, as previously demonstrated (see contour map of sub-catchment B and reach 6). Similar results are showed in case of StSc sensors located in sub-catchment C and different river reaches (third row of Figure 7.5).

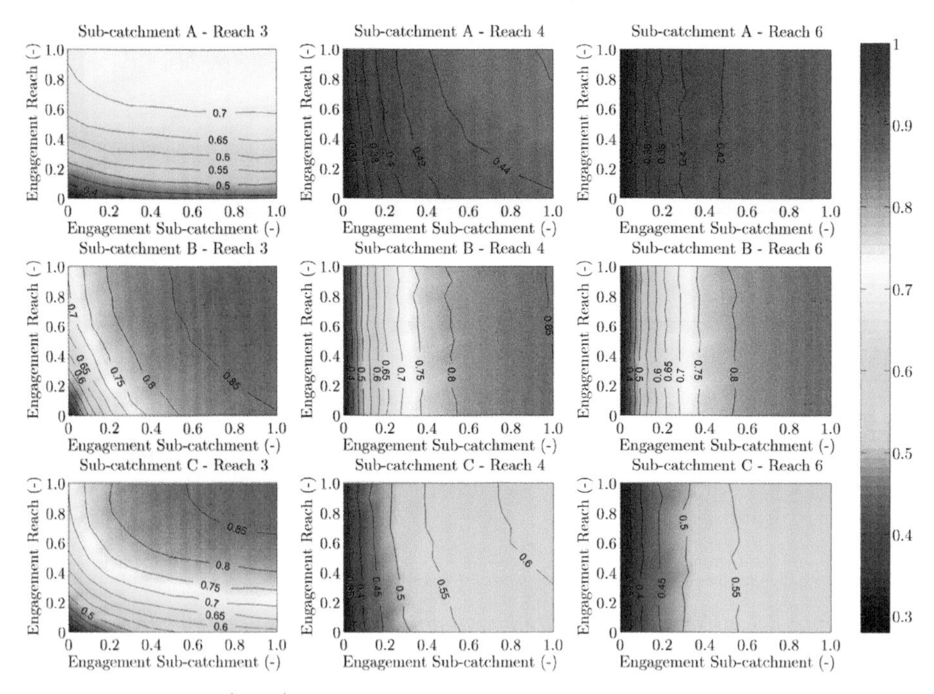

Figure 7.6. μ(NSE) values obtained assimilating CS observations from a combination of StSc sensors located in different sub-catchments and river reaches with 4-hours lead time.

However, engagement levels in upstream river reaches affects the NSE values more than the engagement of StSc sensors in sub-catchment C. Same behaviour is manifested considering StSc sensors located from upstream river reach to downstream.

Third row of Figure 7.5 can be considered as an average situation between first (sub-catchment A) and second (sub-catchment B) row of the same figure. Figure 7.6 is analogous to Figure 7.5 with the only difference that in this case lead time is equal to 4 hours. Overall, NSE values are lower in case of lead time equal to 4 hours than 1 hour, as expected. As previously discussed, assimilation of CS observations in river reaches located upstream the PA station allows to achieve higher NSE values in case of high lead time then StSc located downstream. That is why, also in Figure 7.6 a strong influence of NSE values to different engagement levels of CS observations located in reach 3 can be observed if compared to the ones of CS observations in reach 4 and 6. NSE values are dominated by engagement levels in the sub-catchment A, B and C if compared to the engagement in reach 4 and 6. Intermediate situation is achieved in case of reach 3. In fact, engagement in reach 3 affects the NSE values more than engagement levels in sub-catchment A and C. On the other hand, as in case of Figure 7.5 for 1-hour lead time, engagement in sub-catchment B has higher impact on NSE values than engagement in reach 3.

In Figure 7.7, StSc sensors located in different sub-catchments and river reaches are assimilated at the same time considering three different lead time values. For lead time of 1 hour, high NSE values are achieved even for small engagement values due to the high number of StSc sensors considered in the assimilation process (3 in the sub-catchments and 7 and river reaches). Higher the lead time value, lower the model performances and higher the influence of engagement of the StSc sensors located at the sub-catchment outlet over the sensors located in the river reaches.

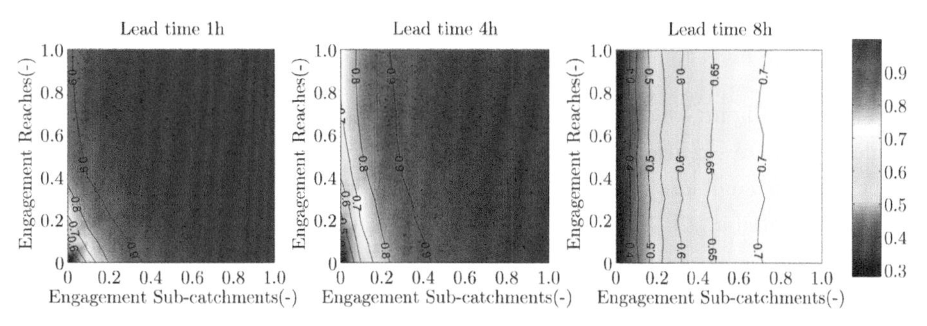

Figure 7.7. μ(NSE) values obtained assimilating CS observations from StSc sensors located in all sub-catchments and river reaches in case of 3 different lead time values

7.4.3 Experiment 7.3

In Experiment 7.3 the effect of assimilating CS observations from DySc sensors is analysed. In this case, the DySc sensors are assumed to be located only along the river reach 3, 4 and 6 so only the hydraulic model is used in this experiment. Also in this case, 100 runs are carried out to account for the random accuracy of CS observations.

Figure 7.8 shows the μ(NSE) values assimilating CS observations from DySc sensors at different location along the three river reaches. In Figure 7.8, for each model run the DySc sensor location is assumed fixed in time. Assimilation from DySc located close to the outlet of the Bacchiglione catchment provides best NSE values in case of engagement equal to one. As expected, NSE values drop for reducing engagement values. Due to the fact that boundary conditions have higher error of the model error, NSE tends to reduce moving from upstream to downstream the given river reach, has demonstrated in section 6.4.2 of the previous chapter.

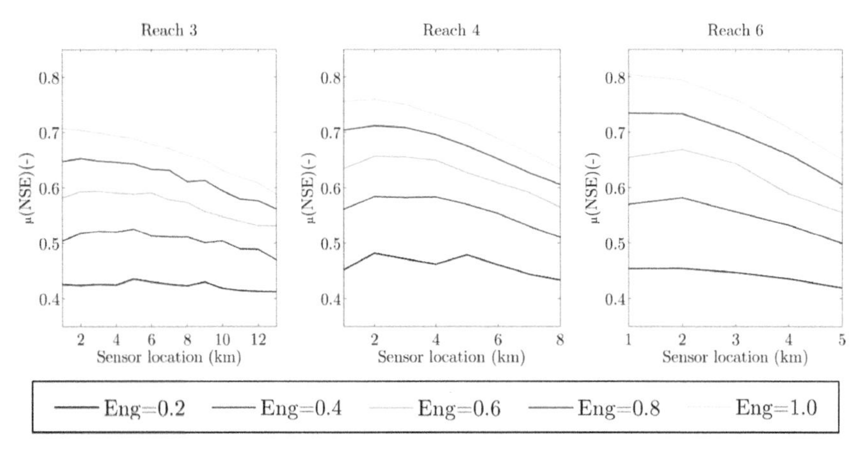

Figure 7.8. Effects of different DySc sensor locations, for fixed engagement levels, on the model performances

As in case of Figure 7.8, also in Figure 7.9 sensor location is assuming fixed in time, while both CS observation accuracy and engagement are variable in time for a given river reach or combination of two of them. However, in this case DySc sensors are assumed located at all the river reach spatial discretization, 1000m, and not at one specific point as in case of Figure 7.8. In most of the cases, μ(NSE) values converge to an asymptotic threshold for increasing engagement levels. Among the three river reaches, 3 and 4 are the ones providing higher NSE values for low engagement levels. This can be related to the high number of DySc sensors located in reach 3 (13 sensors) and 4 (8 sensors).

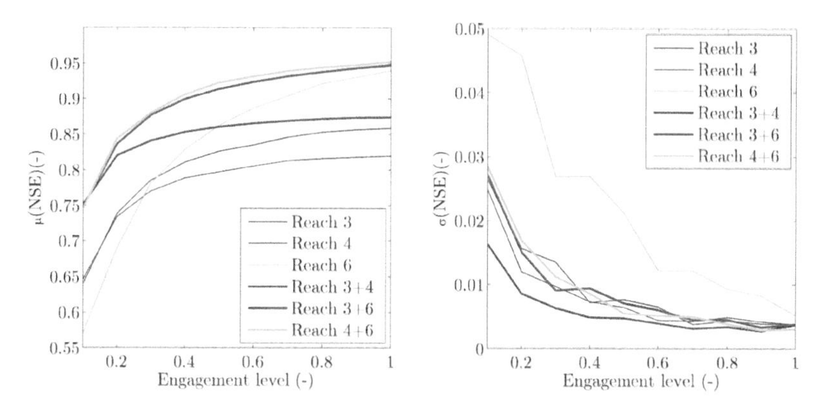

Figure 7.9. Effects of different level of engagement, in terms of μ(NSE) and σ(NSE), in the assimilation of CS observations from DySc sensors

Figure 7.10 represents the μ(NSE) values, functions of different number of sensors and engagement level, obtained considering random location of DySc sensors along the river reaches 3, 4 and 6 in 4 different cases of CS observation bias for 1 hour lead time.

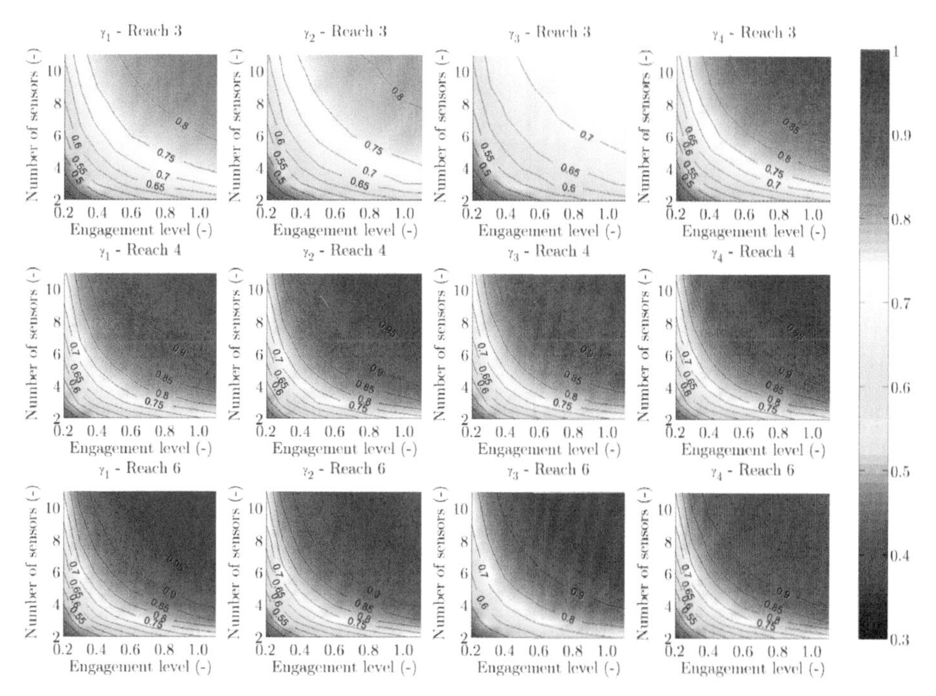

Figure 7.10. μ(NSE) values obtained considering random location of DySc sensors along the river reaches 3, 4 and 6 in 4 different cases of CS observation bias for 1hour lead time

It is worth noting that reach 6 has five different sub-reaches of 1000m. This means that CS observations from only five sensors can be assimilated. However, in Figure 7.10 a total number of 13 DySc sensors is considered. In these experiments, the location of DySc sensors it is randomly generated. It might in fact happen that two sensors are located at distances of 2300m and 2600m from the upstream boundary condition.

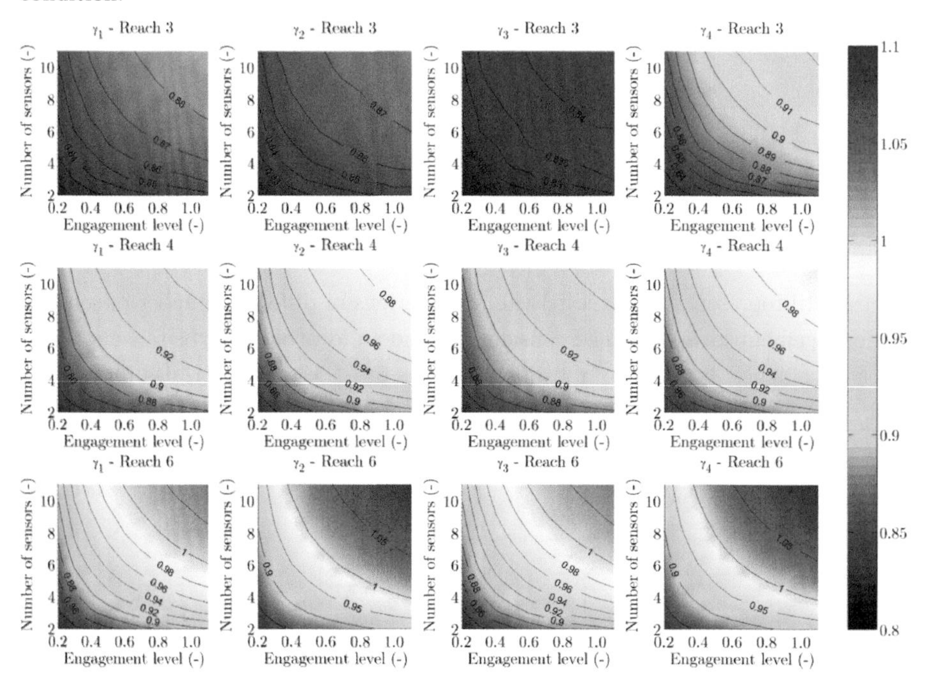

Figure 7.11. Bias *values obtained considering random location of DySc sensors along the river reaches 3, 4 and 6 in 4 different cases of CS observation bias for 1hour lead time*

Because of the small spatial discretization of the hydraulic model, i.e. 1000m, it is assumed that the difference between the hydrographs estimate between two different model discretization it is negligible. For this reason, the two CS observations from the DySc sensors at 2300m and 2600m are simultaneously assimilated at the third sub-reach. In this way, it is possible to assimilate CS observations from a number of DySc sensors higher than the number of model spatial discretization. As it can be observed, different γvalues (bias assumptions) affect the model performances in different ways. Underestimation of the CS observations (γ_3) induces a reduction of the μ(NSE) values due to the underestimated forecasted precipitation which generated a consequent underestimated simulated water depth hydrograph at PA in case of no model updated. For the same reason, overestimation of CS observations ($\gamma 4$) causes an increase in the model performances especially for low number of DySc sensors and engagement level. An intermediate behaviour between γ_3 and γ_4 is obtained in case of γ_2. However, only the indication of the NSE it is not enough to evaluate the obtained results in case of biased observations. For this reason, the estimation of the Bias metric, Eq.(1.3), it is used in Figure 7.11 to provide additional evidence of the results just obtained. Highest Bias values are obtained with DySc located in reach 6 in case of γ_4, while lowest are achieved in reach 3 with γ_3. Reach 4 insures a compromise between high NSE values and Bias metric close to one.

7.4.4 Experiment 7.4

Experiment 7.4 focuses in the assimilation of CS observations from a distributed network of heterogeneous StPh, StSc and DySc sensors. In particular, the engagement level is calculated in a more realistic way accounting for the population living in the surrounding 500m of the river. Six different engagement scenarios are introduced based on three citizen behaviours in collecting and sharing water depth observations. Based on Figure 7.2, different MEL values are calculated.

Figure 7.12 shows μ(NSE) values in case of different engagement scenarios and engagement levels (MEL) according to the different type of sensors. In fact, random value of engagement level between 0 and MEL for the fixed river sub-reach of 1000m is considered for a given model run. In particular, in Figure 7.12, smaller values of MEL such as MEL1, MEL2, MEL3, MEL4 and MEL5 are estimated as to MEL/5, 2MEL/5, 3MEL/5, 4MEL/5 and MEL, respectively. It can be noticed that scenario 1 is the one providing the best model improvements, followed by scenarios 5 and 3. These results demonstrated that sharing CS observations moved by a feeling of belonging to a community of friends (behaviour 2) can help in improving flood prediction if such small community are located upstream a particular target point. The results achieved in case of scenario 3 pointed out that a growing participation, of individualist citizens (behaviour 1), towards sharing hydrological observations in big cities can help to improve model performances. In particular, the model results can benefits to the additional observations provided by weather enthusiasts (behaviour 3). Difference between results obtained with scenario 2 and 3 show the influence of the growth rate parameter in the calculation of the MEL curve in case of same citizen behaviour.

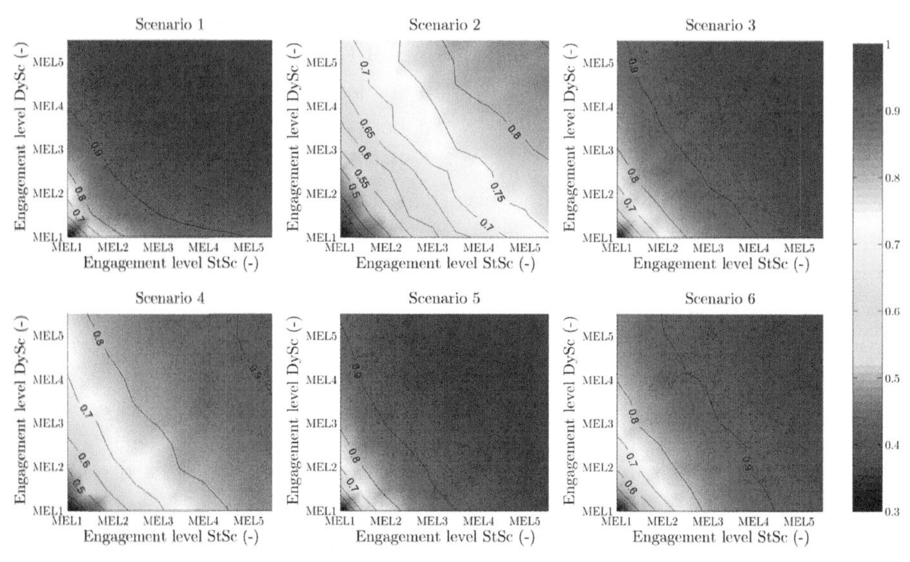

Figure 7.12. μ(NSE) values obtained in case of different engagement scenarios comparing engagement level (MEL) from StSc and DySc sensors

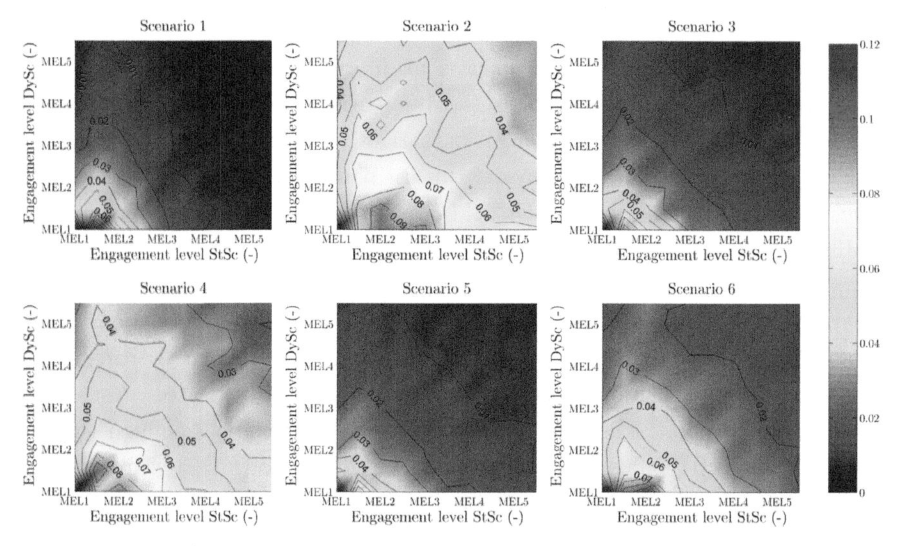

Figure 7.13. σ(NSE) values obtained in case of different engagement scenarios comparing engagement level (MEL) from StSc and DySc sensors

Overall, model results are more sensitive to the change of MEL values in StSc sensors rather than DySc sensors. However, opposite results is showed in scenario 1. It is worth noting that no biased in the CS observations is assumed in case of DySc sensors. Low values of σ(NSE), showed in Figure 7.13, are achieved in scenario 1, 3 and 5.

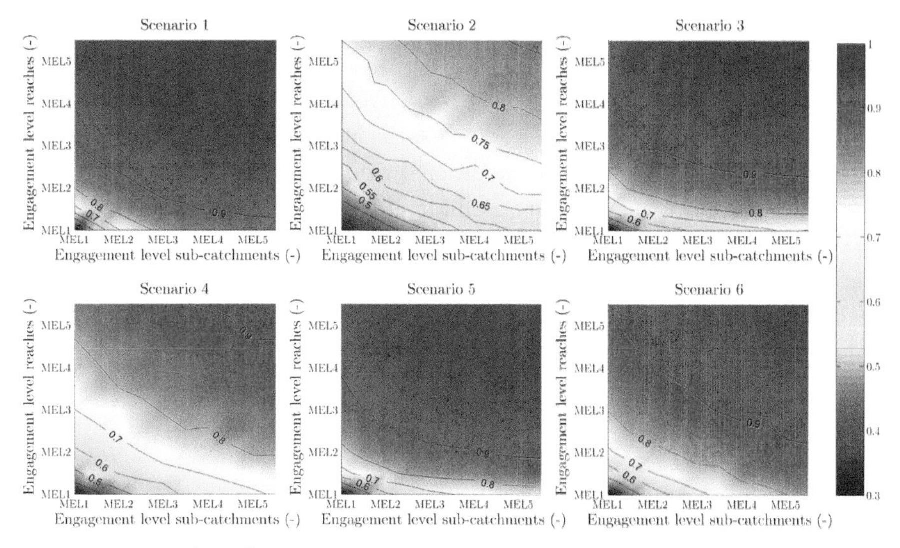

Figure 7.14. μ(NSE) values obtained in case of different engagement scenarios comparing engagement level (MEL) from hydrological (sub-catchments) and hydraulic models (reaches)

Same results can be observed in case of σ(NSE), Figure 7.15.

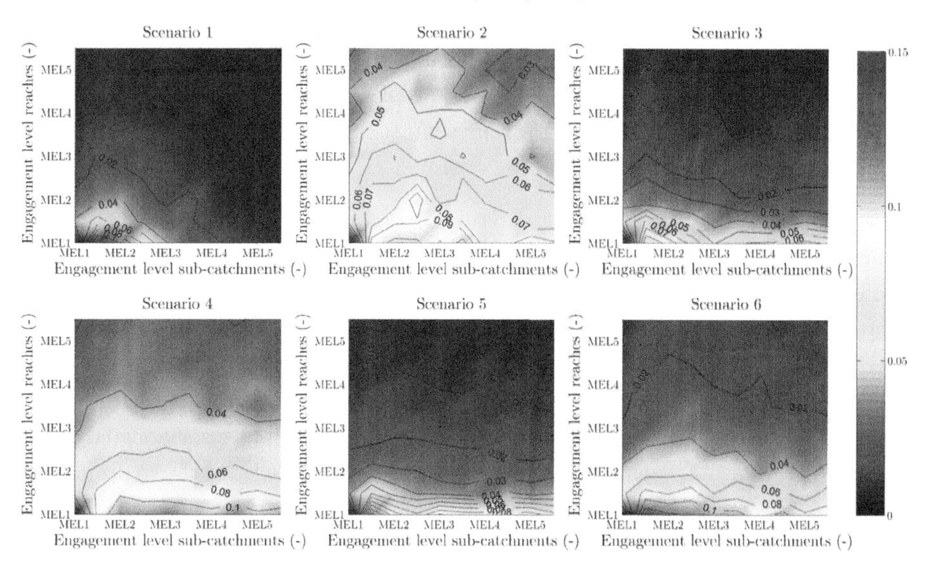

Figure 7.15. σ(NSE) values obtained in case of different engagement scenarios comparing engagement level (MEL) from hydrological (sub-catchments) and hydraulic models (reaches)

It can be observed that low value of σ(NSE) are achieved in scenarios 1, 3, and 5, as showed in Figure 7.13. In addition, variable value of σ(NSE) are obtained in case of different engagement levels along the river reaches, while, no changes in σ(NSE) are visible in case of varying engagement level in the sub-catchments. Figure 7.14 and Figure 7.15 show additional evidence on how CS observations provided by weather enthusiasts are useful in order to increase μ(NSE) values and reduce uncertainty, expressed as σ(NSE), in the model performances. Such results are in fact observed passing from scenario 2 to 4 and then to 6 and from scenario 3 to 5. In the same way, the beneficial effect of a high growth rate in the citizen engagement can be observed moving from scenario 2 to 3 and from scenario 4 to 5.

7.5 CONCLUSIONS

This chapter assessed the effects of assimilating synchronous CS observations provided by citizens by means of a network of distributed heterogeneous sensors on model predictions. In particular, water depth observations from StPh, StSc and DySc sensors, installed within the WeSenseIt Project framework, are assimilated within a semi-distributed hydrological and hydraulic model of the Bacchiglione catchment. Due to the fact that CS water depth observations were not available at the time of writing, synthetic observations having intermittent arrival time and random accuracy in time and space have been used. Four different experiments are carried out. In particular, in experiment 7.4, six citizen engagement levels are introduced in order to give realistic representation of the arrival moment of CS observations.

In Experiment 7.1, streamflow and water depth observations from StPh sensors are assimilated within the hydrological and hydraulic models of the Bacchiglione catchment. Assimilation in the hydrological model provides the best model improvement, in terms of water depth hydrograph at PA (Vicenza), if compared to the other StPh sensors. Assimilation from StPh2 gives a better model prediction, measured by NSE, in case of high lead times, even if for low lead times StPh2 gives a smaller improvement than the assimilation from the StPh3 sensor, located close to the PA station. Assimilation of CS observations at the outlet of the hydrological model provides the best model improvement, also in case of high lead times.

Experiment 7.2 considers only the assimilation of CS observations from StSc sensors. From the results of this experiment it can be shown that, in case of high lead times, assimilation into the hydrological model provides better model predictions, while, for short lead times, assimilation from StSc sensors located in the river reaches close to the target point of PA improves model performance if the sensors are located close to the PA station. NSE values are affected more by the assimilation of CS observations from StSc sensors located in sub-catchment B than sub-catchments A and C. For high lead time values, NSE values are dominated by engagement levels in the sub-catchment A, B and C if compared to the engagement in reaches 4 and 6. Different engagement levels in reach 3 affect the NSE values more than engagement levels in sub-catchments A and C.

Assimilation of CS observations from DySc sensors randomly distributed along the river reaches 3 and 4 is analysed in Experiment 7.3. High values of NSE are achieved for DySc sensors located at points close to the boundary conditions, while when moving the sensor to downstream locations lower NSE values are obtained. This is due to the fact that boundary conditions have higher error than the model error. Different scenarios of observation biases are considered in this experiment. Underestimation of the CS observations induces a reduction of the NSE values due to the underestimated forecasted precipitation which generated a consequent underestimated simulated water depth hydrograph at PA in case of no model update. For the same reason, overestimation of CS observations causes an increase in the model performances especially for a low number of DySc sensors and low engagement level.

The results achieved in Experiment 7.4, where realistic scenarios of engagement levels are proposed, demonstrated that sharing CS observations moved by a feeling of belonging to a community of friends can help in improving flood prediction if the community is located upstream a particular target point. On the other hand, a growing participation, of individualist citizens sharing hydrological observations in big cities can help to improve model performance. In particular, the model results can benefit from the additional observations provided by weather enthusiasts.

8

CONCLUSIONS AND RECOMMENDATIONS

8.1 OVERVIEW

Catastrophic floods cause significant socio-economic losses. Therefore, accurate real-time forecasting of streamflow and water level is crucial for a proper evaluation of the flood risk and subsequent damages. A large number of hydrological and hydraulic models of varying complexity, have been proposed in the last few decades to accurately estimate streamflow and water level along the rivers. Nowadays, model updating techniques, in particular data assimilation methods, have been used to improve flood forecasts by integrating static ground observations, and in some cases also remote sensing observations, within hydrological and hydrodynamic models. In recent years, continued technological improvement has stimulated the spread of low-cost sensors that allow for employing crowdsourced data and obtaining observations of hydrological variables in a more distributed way than the classical static physical sensors. However, current hydrologic and hydraulic research typically considers assimilation of observations coming from traditional static sensors. One reason for this is that crowdsourced measurements have random arrival frequency and varying accuracy. This PhD research aims to develop and test the methods for assimilating CS observations having variable spatio-temporal coverage and provided by citizens by means of different low-cost sensors, and to demonstrate that such data can be useful for improving flood forecasts. The proposed methods have been successfully implemented in the AMICO EWS in the

Bacchiglione catchment, an official case study of the WeSenseIt EU project which is funding this PhD research.

8.2 RESEARCH OUTCOMES

The results of this study demonstrate that crowdsourced observations can significantly improve flood prediction if integrated into hydrological and hydraulic models. In particular, this study points out that networks of low-cost static social (StSc) and dynamic social (DySc) sensors can complement traditional networks of static physical (StPh) sensors and improve the accuracy of flood forecasting. In addition, citizen engagement levels play a fundamental role in the assimilation of CS observations and consequent improvement of flood prediction. This research can contribute to the recent efforts to build citizens' water observatories, which can make citizens an active part of information capturing, evaluation and communication, helping at the same time to improve model-based flood forecasting. Below, the research outcomes in relation to the specific research objectives introduced in section 1.4 are reported.

1. *To investigate the effect of StPh sensor locations and different observation accuracies on the assimilation of distributed synchronous streamflow observations in hydrological modelling.*

The results of this PhD research have proved that assimilation of streamflow observations from StPh sensors provides improvements in model performance, the magnitude of which depends on the location of such observations and the structure of a semi-distributed hydrological model (Brue catchment case study). In fact, the best model improvement are achieved by assimilation of streamflow observations within the whole catchment or along the main river reach. In this last case, sensor locations which generate the highest NSE value at the catchment outlet change according to the given flood event. This may be due to the fact that different flood events are generated from a different spatial distribution of precipitation. Therefore, designing flow sensor networks considering a longer time series of flood events is not optimal since the effect of the single flood event is not considered. For this reason, additional study towards the development of techniques to design networks of dynamic low-cost sensors should be carried out.

Flood forecasts are influenced by the total number of StPh sensors and their locations in the case of model structure MS2, while results achieved in the case of MS1 are more sensitive to the locations of the StPh sensors but not to their number.

Overall, assimilation of streamflow observations provided comparable results in terms of forecasting accuracy for the two model structures. However, for high lead time values, MS1 is generally better than MS2. These findings can be used as criteria to develop methods for streamflow monitoring network design. Regarding the influence of observation accuracy error on the DA performances, this research indicated that no significant effect on the outflow hydrograph prediction is obtained considering the error in biased streamflow observations (ErrSD) and error in the water level measurements (ErrWL) if compared to the results obtained assuming rating curve uncertainty (ErrRC). Moreover, the effect of the different types of observational errors is observed only for low lead times. MS2 is more sensitive than MS1 to different types of observational errors for low lead time values.

2. *To investigate the effect of assimilating distributed uncertain synchronous CS streamflow observations, intermittent in time and space, from StSc sensors in hydrological modelling.*

This PhD research demonstrates that assimilation of CS streamflow observations at interior points of the Brue catchment can improve the hydrologic models depending on the particular location of the StSc sensors and hydrological model structures. Lower accuracy, variable in time and space, is assumed for CS data from StSc sensors compared to that of StPh sensors. As in the case of StPh sensors, the structure of the semi-distributed model and the locations of sensors affect the assimilation of streamflow observations from StSc sensors in different ways. MS1 is less sensitive than MS2 to the assimilation of observations from StSc, while both model structures are influenced by the assimilation of intermittent observations. Realistic assumptions about the spatio-temporal configuration of CS streamflow observations in the case of a heterogeneous network of StPh, optimally and non-optimally located, and StSc sensors are introduced. Such configurations assume reasonable locations, next to urban areas where citizens can provide data, and temporal availability, mainly daylight hours, of CS observations from StSc sensors. Integration of CS streamflow measurements from StSc sensors during peak flow hours, which can be carried out by trained volunteers in the contexts of citizen observatories, allows for further improvements in model accuracy, comparable to those obtained by assimilating the observations from an optimal network of StPh sensors only. Assimilation of CS data from StSc sensors provides similar results to assimilation of streamflow observations from a non-optimal network of StPh sensors, leading to the conclusion that an inappropriate distribution of StPh sensors can be replaced by distributed StSc sensors. Finally, it is demonstrated that a non-optimal network of StPh sensors can be integrated with a network of StSc sensors providing intermittent CS observations only during daylight and peak hours, in order to improve model performance.

3. To assess the influence of assimilating asynchronous CS streamflow observations from StSc sensors in hydrological modelling.

Citizen-based CS observations are generally characterised by random accuracy and are derived at random moments, which may not coincide with the model time step. In this PhD thesis it is found that assimilation of asynchronous observations results in a significant improvement of NSE for different lead time values. Such analysis is performed for a lumped hydrological model for the Brue catchment. Increasing the number of assimilated CS asynchronous observations within two model time steps induces an improvement in the NSE. However, after a threshold number of CS observations NSE asymptotically approaches a certain value meaning that no improvement is achieved with additional observations. Accuracy of CS observations influences the NSE values more than the arrival moments of such data. NSE values drop when the intervals between the assimilated observations are too large. In this way the abundance of CS data is no longer able to compensate for their intermittency.

In experiments with the Bacchiglione catchment, in order to improve model performances, the distributed StSc sensors providing asynchronous CS observations are integrated with a single StPh sensor. In this case, a semi-distributed hydrological-hydraulic model is used to forecast streamflow and water depth at the outlet of the Bacchiglione catchment. The results demonstrate that networks of low-cost StSc sensors can complement traditional networks of StPh sensors and improve the accuracy of flood forecasting even in cases of a small number of intermittent asynchronous CS observations. As expected, the replacement of a StPh sensor for a StSc sensor at only one location does not improve the model performance in terms of NSE for different lead time values.

4. To study the effects of different DA approaches in the assimilation of synchronous streamflow observations, from existing StPh sensors, in hydraulic modelling.

The effects of different DA approaches, such as DI, NS, KF, EnKF and AEnKF, are studied in relation to the assimilation of streamflow observations from StPh sensors in lumped and distributed structures of the hydraulic model implemented for the Trinity and Sabine Rivers. Overall, assimilation of streamflow observations in both model structures increases model performance. DI and AEnKF provide the best model improvement in the case of the lumped model. KF and EnKF seem to be extremely sensitive to the proper definition of model error. On the other hand, in the case of the distributed model, DI does not give satisfactory model improvement if compared to that achieved using AEnKF. In fact, using DI and NS, model states are updated only at the assimilation location, while in the case of Kalman filtering approaches, the update is performed along the whole river reach because of the distributed nature of the Kalman gain and covariance matrix. Because of the nature of the model, the update is only effective in the model states downstream of the assimilation point in all the considered DA. Kalman filtering methods allow for a smooth update of the model state along the river reach, while in the case of DI and NS, abrupt changes of the model state at the assimilation point occur. Increasing the number of past observations in AEnKF increases the model performances expressed in terms of NSE, R and Bias. Ensemble methods showed variable water profiles upstream of the assimilation point. This effect can be reduced by locating an additional sensor close to the boundary conditions in order to reduce the ensemble spread and hence model uncertainty.

5. *To evaluate the effects of assimilation of synchronous water depth observations from spatially distributed StPh sensors in hydraulic modelling.*

The results obtained in this PhD research point out that assimilation of water depth observations from StPh sensors located along the Bacchiglione River can improve flood predictions. In particular, KF is noticeably sensitive to model error and sensor location. High error in the boundary condition tends to better improve water profile when the assimilation point is closer to it and a smooth update is achieved downstream of the assimilation point. On the other hand, in cases when the model error is higher than the boundary error, an abrupt update is obtained at the water depth sensor location and good model performances are achieved if the StPh sensor is located close to the reach outlet. For this reason, in cases where the model error is higher than the boundary error, it is suggested that the sensor be located downstream of the river reach to maximize the model improvement at the river outlet. Assimilation of water depth observations from StPh sensors located in reaches 3, 4 and 6 of the Bacchiglione River gives additional improvement to the model results at the outlet if compared to the model with no update. However, reach 6 tends to lose the assimilation effects faster than reaches 3 and 4, in the case of flood prediction, due to its shorter travel time. In fact, StPh sensors located at the upstream part of such reaches ensure additional lead time, up to 6 hours, for the prediction of water depth at the reach outlet. For this reason, the choice of the optimal location of StPh sensors should be a compromise between best NSE value and prediction capability of the model itself.

6. *To assess the integration of distributed StPh, StSc and DySc sensors for assimilation of synchronous CS observations within a cascade of hydrological and hydraulic models.*

Water depth observations from StPh, StSc and DySc sensors, installed within the WeSenseIt Project framework, are assimilated within the semi-distributed hydrological and hydraulic model of the Bacchiglione catchment. Different model runs are performed in order to account for the random accuracy and engagement level of the citizen providing CS observations. Assimilation of CS observations in river reaches close to the catchment outlet guarantees the best model performances. However, this result is valid only in the case of low lead time values. In fact, assimilation of CS observations in hydrological modesl ensures the best model improvement, especially in the case of high lead time values. NSE values are more affected by the assimilation of CS observations from StSc sensors located in sub-catchment B rather than those in sub-catchments A and C. In terms of citizens' actions, one can say that different engagement levels in reach 3 affects the NSE values more than engagement levels in sub-catchments A and C. Due to the fact that boundary conditions have higher model error, high values of NSE are achieved for DySc sensors located at points close to the boundary conditions, while moving the sensor at downstream locations lower NSE values are obtained. Different observation biases are also considered. Because of the underestimated simulated water depth hydrograph at PA, an underestimation of the assimilated CS observations induces a reduction of the NSE values. In the same way, overestimation of CS observations generates a model improvement especially for low number of DySc sensors and engagement level. In the case of realistic scenarios of engagement levels, this PhD research demonstrates that sharing CS observations (e.g. being motivated by a feeling of belonging to a community) can help in improving flood prediction if such a small community is located upstream of a particular target point. On the other hand, growing participation of citizens towards sharing hydrological observations in large cities can help to improve model performances. In particular, the model results can benefit from the additional observations provided by weather enthusiasts.

7. To develop guidelines for using technologies for crowdsourced data assimilation in flood forecasting

The choice of the proper mathematical model and updating technique to be used for flood forecasting may vary according to the data availability, location of the sensors, type of forecast, etc. In the case of crowdsourced observations, the type of data, their location and the location of the area in which the forecast has to be performed (i.e. the target area, which can be Vicenza in the case of the Bacchiglione catchment) are extremely important for assessing the type of mathematical model to be used. Figure 8.1 shows a decision tree that can be used to select the proper model and updating technique based on the results achieved in this thesis. The first important point is which type of sensors providing crowdsourced observations are used within the given catchment. Such sensors might be dynamic or static. In the case of dynamic sensors, as mentioned in the previous chapters, the observations might arrive at random positions, which leads to the use of a semi-distributed or distributed model in order to be able to represent the spatial variability of such observations. In the case of static sensors, the randomness is related to the arrival moment (and uncertainty) of the crowdsourced observations and the choice of the proper model is connected to the location of the sensor and the forecast location or target area. In fact, where the only sensors and the target area are located at the outlet of the catchment, as in the Experiment 5.1, a conceptual lumped model can be used for flood forecasting. On the other hand, where the sensor is located at the outlet but the target area is inside the catchment, as in chapter 4 where towns are located upstream of the outlet section of the Brue catchment, a semi-distributed/distributed model may be more suitable. In the extreme case of static social sensors distributed within the catchment, as in the case of the Bacchiglione catchment, a semi-distributed/distributed model is suggested. Finally, according to the characteristics of the model and observation uncertainty different model updating techniques can be used to assimilate crowdsourced observations. IIn the case of a linear model and normal observational error, a Kalman filter based approach can be used. On the other hand, for a non-linear model and normal observational error, ensemble methods can be used. In the situation where there is both a non-linear model and non-Gaussian observation error a particle filter or variational methods might be used.

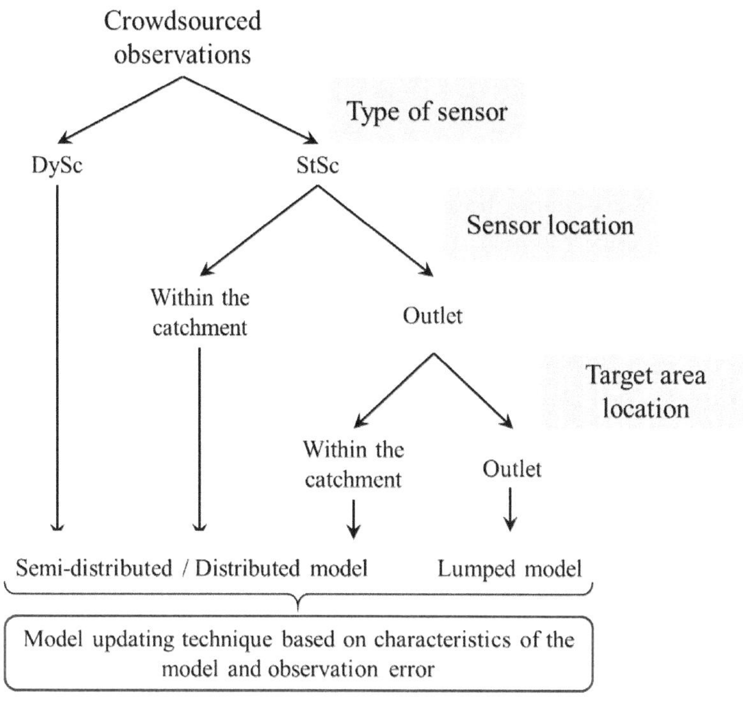

Figure 8.1. Decision tree used to select the appropriate mathematical model and updating technique based on type of crowdsourced observations, sensor location and target area

8.3 LIMITATIONS AND RECOMMENDATIONS

This work still has certain limitations which should be mentioned, and which lead to formulation of some recommendations for future work.

The spatial discretisation of the semi-distributed model implemented in the Brue catchment (see chapter 2), with a drainage area about 2 km^2, may not be small enough to assume that any CS observation is actually occurring at the corresponding sub-catchment's outlet. To account for that, a fully distributed hydrological model should be implemented in order to test and validate the obtained results.

The proposed DA methods used to assimilate CS observations are applied to the linear parts of hydrological and hydraulic models. For this reason, the next steps in this research will be the implementation of the proposed methodology to non-linear hydrological and hydraulic models. In particular, the spatial and temporal correlation within the model error matrix should be included in future studies.

Additional analyses with different case studies, having different topographic and hydrological characteristics from the ones considered in this research, and longer time series of flood events (due to the limited flood events considered in this research) should be carried out. This will permit drawing more general conclusions about assimilation of CS observations and their additional value in different types of catchments. For example, because of practical limitations for sensor locations (e.g. sensors are usually located at bridges where flow disturbance is minimal), our finding that upstream locations could have more benefits may not work for other rivers.

Due to the lack of real-time CS observations in the case studies considered, synthetic streamflow observations, having some assumed frequency and accuracy characteristics (albeit quite realistic), are used. For this reason, to further validate the results obtained in this research, real-life CS observations provided by citizens using StSc and DySc sensors should be used.

Accuracy of CS observations is considered in this work to be a random variable. However, advancing methods for a more accurate assessment of the data quality and accuracy of streamflow and water depth CS observations coming from StSc and DySc sensors need to be considered. In this way, based on the CS observations' accuracy, a pre-filtering module aiming to select only observations having good accuracy while discarding the ones with low accuracy, might be used.

Regarding the observation error in the DA performances, the value of the coefficient α of Eq.(4.4, in the case of ErrRC+WL, is assumed equal to the sum of the two previous coefficients α_{RC} and α_{WL}. Additional studies should be carried out to consider the combined error in both measurement and uncertain rating curve as a joint probability distribution and not just as a normal distribution.

Finally, no specific spatial sensor trajectory of the citizens moving from one StSc sensor to another or using DySc sensors is considered since this would require the introduction of assumptions about citizens' behaviour in the case of a flood event. This component would be extremely important in the case of dynamic sensors but it could not be included in this research due to the lack of information about citizen engagement in monitoring river water level in all the case studies. That is why Agent Based Models (ABM), which simulate the interactions between autonomous agents, could be introduced in future studies and integrated into the presented mathematical modelling framework.

REFERENCES

Abbott, M. B. and Ionescu, F.: On The Numerical Computation Of Nearly Horizontal Flows, J. Hydraul. Res., 5(2), 97–117, doi:10.1080/00221686709500195, 1967.

ABC: ABC's crowdsourced flood-mapping initiative, ABCs Crowdsourced Flood-Mapp. Initiat. [online] http://www.abc.net.au/technology/articles/2011/01/13/3112261.htm (Accessed 20 January 2016), 2011.

Abebe, A. and Price, R.: Information Theory and Neural Networks for Managing Uncertainty in Flood Routing, J. Comput. Civ. Eng., 18(4), 373–380, doi:10.1061/(ASCE)0887-3801(2004)18:4(373), 2004.

Aerts, J. C. J. H., Botzen, W. J. W., Emanuel, K., Lin, N., Moel, H. de and Michel-Kerjan, E. O.: Evaluating Flood Resilience Strategies for Coastal Megacities, Science, 344(6183), 473–475, doi:10.1126/science.1248222, 2014.

Alberoni, P., Collier, C. and Khabiti, R.: ACTIF Best practice paper - Understanding and reducing uncertainty in flood forecasting, Proceeding Act. Conf., (1), 1–43, 2005.

Alfonso, L.: Use of hydroinformatics technologies for real time water quality management and operation of distribution networks. Case study of Villavicencio, Colombia, M.Sc. Thesis, UNESCO-IHE, Institute for Water Education, Delft, The Netherlands., 2006.

Alfonso, L. and Price, R.: Coupling hydrodynamic models and value of information for designing stage monitoring networks, Water Resour. Res., 48(8), W08530, doi:10.1029/2012WR012040, 2012.

Alfonso, L. and Tefferi, M.: Effects of Uncertain Control in Transport of Water in a River-Wetland System of the Low Magdalena River, Colombia, in Transport of Water versus Transport over Water, edited by C. Ocampo-Martinez and R. R. Negenborn, pp. 131–144, Springer International Publishing. [online] Available from: http://link.springer.com/chapter/10.1007/978-3-319-16133-4_8 (Accessed 6 May 2016), 2015.

Alfonso, L., Lobbrecht, A. and Price, R.: Optimization of water level monitoring network in polder systems using information theory, Water Resour. Res., 46(12), W12553, doi:10.1029/2009WR008953, 2010.

Alfonso, L., He, L., Lobbrecht, A. and Price, R.: Information theory applied to evaluate the discharge monitoring network of the Magdalena River, J. Hydroinformatics, 15(1), 211, doi:10.2166/hydro.2012.066, 2013.

Alfonso, L., Mukolwe, M. M. and Di Baldassarre, G.: Probabilistic flood maps to support decision-making: Mapping the Value of Information, Water Resour. Res., doi:10.1002/2015WR017378, 2016.

van Andel, S. J., Weerts, A., Schaake, J. and Bogner, K.: Post-processing hydrological ensemble predictions intercomparison experiment, Hydrol. Process., 27(1), 158–161, doi:10.1002/hyp.9595, 2013.

Anderson, J. L.: An Ensemble Adjustment Kalman Filter for Data Assimilation, Mon. Weather Rev., 129(12), 2884–2903, doi:10.1175/1520-0493(2001)129<2884:AEAKFF>2.0.CO;2, 2001.

Andreadis, K., Das, N., Granger, S., Han, E., Ines, A. and Stampoulis, D.: Assimilating multi-sensor satellite observations for initializing hydrologic and agricultural forecasts, vol. 17, p. 7705. [online] Available from: http://adsabs.harvard.edu/abs/2015EGUGA..17.7705A (Accessed 18 March 2016), 2015.

Andreadis, K. M. and Lettenmaier, D. P.: Assimilating remotely sensed snow observations into a macroscale hydrology model, Adv. Water Resour., 29(6), 872–886, doi:10.1016/j.advwatres.2005.08.004, 2006.

Andreadis, K. M. and Schumann, G. J.-P.: Estimating the impact of satellite observations on the predictability of large-scale hydraulic models, Adv. Water Resour., 73, 44–54, doi:10.1016/j.advwatres.2014.06.006, 2014.

Andreadis, K. M., Clark, E. A., Lettenmaier, D. P. and Alsdorf, D. E.: Prospects for river discharge and depth estimation through assimilation of swath-altimetry into a raster-based hydrodynamics model, Geophys. Res. Lett., 34(10), doi:10.1029/2007GL029721, 2007.

Arnold, C. P. and Dey, C. H.: Observing-Systems Simulation Experiments: Past, Present, and Future, Bull. Am. Meteorol. Soc., 67(6), 687–695, doi:10.1175/1520-0477(1986)067<0687:OSSEPP>2.0.CO;2, 1986.

Aronica, G., Hankin, B. and Beven, K.: Uncertainty and equifinality in calibrating distributed roughness coefficients in a flood propagation model with limited data, Adv. Water Resour., 22(4), 349–365, doi:10.1016/S0309-1708(98)00017-7, 1998.

Arulampalam, M. S., Maskell, S., Gordon, N. and Clapp, T.: A tutorial on particle filters for online nonlinear/non-Gaussian Bayesian tracking, Signal Process. IEEE Trans. On, 50(2), 174–188, 2002.

Au, J., Bagchi, P., Chen, B., Martinez, R., Dudley, S. A. and Sorger, G. J.: Methodology for public monitoring of total coliforms, Escherichia coli and toxicity in waterways by Canadian high school students, J. Environ. Manage., 58(3), 213–230, doi:10.1006/jema.2000.0323, 2000.

Aubert, D., Loumagne, C. and Oudin, L.: Sequential assimilation of soil moisture and streamflow data in a conceptual rainfall–runoff model, J. Hydrol., 280(1–4), 145–161, doi:10.1016/S0022-1694(03)00229-4, 2003.

Aureli, F., Mignosa, P., Ziveri, C. and Maranzoni, A.: Fully-2D and quasi-2D modeling of flooding scenarios due to embankment failure, River Flow 2006 - Ferreira Alves Leal Cardoso Eds © 2006 Taylor Francis Group Lond., 1473–1482, doi:10.1201/9781439833865.ch157, 2006.

Auroux, D., Bansart, P. and Blum, J.: An easy-to-implement and efficient data assimilation method for the identification of the initial condition: the Back and Forth Nudging (BFN) algorithm, J. Phys. Conf. Ser., 135(1), 12011, doi:10.1088/1742-6596/135/1/012011, 2008.

Babovic, V., Canizares, R., Jensen, H. R. and Kinting, A.: Neural Networks as Routine for Error Updating of Numerical Models, J. Hydraul. Eng., 127(3), 181–193, doi:10.1061/(ASCE)0733-9429(2001)127:3(181), 2001.

Bardossy, A., Bogardi, I. and Duckstein, L.: Fuzzy regression in hydrology, Water Resour. Res., 26(7), 1497–1508, doi:10.1029/WR026i007p01497, 1990.

Bardossy, A., Duckstein, L. and Bogardi, I.: Fuzzy rule-based classification of atmospheric circulation patterns, Int. J. Climatol., 15(10), 1087–1097, doi:10.1002/joc.3370151003, 1995.

Barth, B., Ringen, S. and Sallas, J.: City of Dallas Floodway System (DFS) Case Study: 100-Year Levee Remediation, in Rocky Mountain Geo-Conference 2014, pp. 59–68, American Society of Civil Engineers, 2014.

Bartholmes, J. and Todini, E.: Coupling meteorological and hydrological models for flood forecasting, Hydrol. Earth Syst. Sci. Discuss., 9(4), 333–346, 2005.

Bates, P. D. and De Roo, A. P. J.: A simple raster-based model for flood inundation simulation, J. Hydrol., 236(1–2), 54–77, doi:10.1016/S0022-1694(00)00278-X, 2000.

Batson, C. D., Ahmad, N. and Tsang, J.-A.: Four motives for community involvement, J. Soc. Issues, 58(3), 429–445, doi:10.1111/1540-4560.00269, 2002.

van Beek, L. P. H., Wada, Y. and Bierkens, M. F. P.: Global monthly water stress: 1. Water balance and water availability, Water Resour. Res., 47(7), W07517, doi:10.1029/2010WR009791, 2011.

Bergström, S.: Principles and confidence in hydrological modelling, Hydrol. Res., 22(2), 123–136, 1991.

Beven, K.: How far can we go in distributed hydrological modelling?, Hydrol. Earth Syst. Sci., 5(1), 1–12, doi:10.5194/hess-5-1-2001, 2001.

Beven, K. and Binley, A.: The future of distributed models: Model calibration and uncertainty prediction, Hydrol. Process., 6(3), 279–298, doi:10.1002/hyp.3360060305, 1992.

Biancamaria, S., Durand, M., Andreadis, K. M., Bates, P. D., Boone, A., Mognard, N. M., Rodríguez, E., Alsdorf, D. E., Lettenmaier, D. P. and Clark, E. A.: Assimilation of virtual wide swath altimetry to improve Arctic river modeling, Remote Sens. Environ., 115(2), 373–381, doi:10.1016/j.rse.2010.09.008, 2011.

Bird, T. J., Bates, A. E., Lefcheck, J. S., Hill, N. A., Thomson, R. J., Edgar, G. J., Stuart-Smith, R. D., Wotherspoon, S., Krkosek, M., Stuart-Smith, J. F., Pecl, G. T., Barrett, N. and Frusher, S.: Statistical solutions for error and bias in global citizen science datasets, Biol. Conserv., 173, 144–154, doi:10.1016/j.biocon.2013.07.037, 2014.

Blöschl, G., Reszler, C. and Komma, J.: A spatially distributed flash flood forecasting model, Environ. Model. Softw., 23(4), 464–478, doi:10.1016/j.envsoft.2007.06.010, 2008.

Bonney, R., Ballard, H., Jordan, R., McCallie, E., Phillips, T., Shirk, J. and Wilderman, C. C.: Public Participation in Scientific Research: Defining the Field and Assessing Its Potential for Informal Science Education. A CAISE Inquiry Group Report., 2009.

Bonney, R., Shirk, J. L., Phillips, T. B., Wiggins, A., Ballard, H. L., Miller-Rushing, A. J. and Parrish, J. K.: Next Steps for Citizen Science, Science, 343(6178), 1436–1437, doi:10.1126/science.1251554, 2014.

Bordogna, G., Carrara, P., Criscuolo, L., Pepe, M. and Rampini, A.: A linguistic decision making approach to assess the quality of volunteer geographic information for citizen science, Inf. Sci., 258, 312–327, doi:10.1016/j.ins.2013.07.013, 2014.

Box, G. E. P., Jenkins, G. M. and Reinsel, G. C.: Time Series Analysis: Forecasting and Control, 4th Edition., San Francisco. [online] Available from: http://eu.wiley.com/WileyCDA/WileyTitle/productCd-0470272848.html (Accessed 10 February 2016), 1970.

Brandimarte, L. and Baldassarre, G. D.: Uncertainty in design flood profiles derived by hydraulic modelling, Hydrol. Res., 43(6), 753–761, doi:10.2166/nh.2011.086, 2012.

Brandimarte, L. and Woldeyes, M. K.: Uncertainty in the estimation of backwater effects at bridge crossings, Hydrol. Process., 27(9), 1292–1300, doi:10.1002/hyp.9350, 2013.

Brocca, L., Melone, F., Moramarco, T., Wagner, W., Naeimi, V., Bartalis, Z. and Hasenauer, S.: Improving runoff prediction through the assimilation of the ASCAT soil moisture product, Hydrol. Earth Syst. Sci., 14(10), 1881–1893, doi:10.5194/hess-14-1881-2010, 2010.

Brocca, L., Moramarco, T., Melone, F., Wagner, W., Hasenauer, S. and Hahn, S.: Assimilation of surface- and root-zone ASCAT soil moisture products into rainfall-runoff modeling, IEEE Trans. Geosci. Remote Sens., 50(7 PART1), 2542–2555, doi:10.1109/TGRS.2011.2177468, 2012.

Buizza, R.: The value of probabilistic prediction, Atmospheric Sci. Lett., 9(2), 36–42, doi:10.1002/asl.170, 2008.

Burnash, R. J. C.: The NWS River Forecast System - catchment modeling., 311–366, 1995.

Buytaert, W., Zulkafli, Z., Grainger, S., Acosta, L., Alemie, T. C., Bastiaensen, J., De BiÃ¨vre, B., Bhusal, J., Clark, J., Dewulf, A., Foggin, M., Hannah, D. M., Hergarten, C., Isaeva, A., Karpouzoglou, T., Pandeya, B., Paudel, D., Sharma, K., Steenhuis, T., Tilahun, S., Van Hecken, G. and Zhumanova, M.: Citizen science in hydrology and water resources: opportunities for knowledge generation, ecosystem service management, and sustainable development, Front. Earth Sci., 2(October), 1–21, doi:10.3389/feart.2014.00026, 2014.

Canizares, R., Heemink, A. W. and Vested, H. J.: Application of advanced data assimilation methods for the initialisation of storm surge models, J. Hydraul. Res., 36(4), 655–674, doi:10.1080/00221689809498614, 1998.

Cao, W. Z., Bowden, W. B., Davie, T. and Fenemor, A.: Multi-variable and multi-site calibration and validation of SWAT in a large mountainous catchment with high spatial variability, Hydrol. Process., 20(5), 1057–1073, doi:10.1002/hyp.5933, 2006.

Castell, N., Kobernus, M., Liu, H.-Y., Schneider, P., Lahoz, W., Berre, A. J. and Noll, J.: Mobile technologies and services for environmental monitoring: The Citi-Sense-MOB approach, Urban Clim., 14, Part 3, 370–382, doi:10.1016/j.uclim.2014.08.002, 2015.

Célleri, R., Buytaert, W., De Bièvre, B., Tobón, C., Crespo, P., Molina, J. and Feyen, J.: Understanding the hydrology of tropical Andean ecosystems through an Andean Network of Basins, DOI: 10.13140/2.1.4187.3608, 2009.

Chen, J., Zhang, W., Gao, J. and Cao, K.: Assimilating multi-site measurements for semi-distributed hydrological model updating, Quat. Int., 282, 122–129, doi:10.1016/j.quaint.2012.01.030, 2012.

Chen, M., Liu, S., Tieszen, L. L. and Hollinger, D. Y.: An improved state-parameter analysis of ecosystem models using data assimilation, Ecol. Model., 219(3–4), 317–326, doi:10.1016/j.ecolmodel.2008.07.013, 2008.

Cifelli, R., Doesken, N., Kennedy, P., Carey, L. D., Rutledge, S. A., Gimmestad, C. and Depue, T.: The Community Collaborative Rain, Hail, and Snow Network: Informal Education for Scientists and Citizens, Bull. Am. Meteorol. Soc., 86(8), 1069–1077, 2005.

Cipra, T. and Romera, R.: Kalman filter with outliers and missing observations, Test, 6(2), 379–395, doi:10.1007/BF02564705, 1997.

Ciravegna, F., Huwald, H., Lanfranchi, V. and Wehn de Montalvo, U.: Citizen observatories: the WeSenseIt vision, Florence, Italy., 2013.

Clark, M. P., Rupp, D. E., Woods, R. A., Zheng, X., Ibbitt, R. P., Slater, A. G., Schmidt, J. and Uddstrom, M. J.: Hydrological data assimilation with the ensemble

Kalman filter: Use of streamflow observations to update states in a distributed hydrological model, Adv. Water Resour., 31(10), 1309–1324, doi:10.1016/j.advwatres.2008.06.005, 2008.

Cooperman, R. L.: Tactical Ballistic Missile Tracking using the Interacting Multiple Model Algorithm, 5th Int. Conf. Inf. Fusion, 2, 824–831, doi:10.1109/ICIF.2002.1020892, 2002.

Corine Land Cover: Corine Land Cover 2006 raster data — European Environment Agency, [online] Available from: http://www.eea.europa.eu/data-and-maps/data/corine-land-cover-2006-raster (Accessed 20 March 2016), 2006.

Cortes Arevalo, V. J.: Use of volunteers' information to support proactive inspection of hydraulic structures, PhD Thesis, TU Delft, Delft, The Netherlands, 2016.

Cortes Arevalo, V. J., Charrière, M., Bossi, G., Frigerio, S., Schenato, L., Bogaard, T., Bianchizza, C., Pasuto, A. and Sterlacchini, S.: Evaluating data quality collected by volunteers for first-level inspection of hydraulic structures in mountain catchments, Nat. Hazards Earth Syst. Sci., 14(10), 2681–2698, doi:10.5194/nhess-14-2681-2014, 2014.

Corzo, G. and Solomatine, D.: Baseflow separation techniques for modular artificial neural network modelling in flow forecasting, Hydrol. Sci. J., 52(3), 491–507, 2007.

Cunge, J. A.: On The Subject Of A Flood Propagation Computation Method (Muskingum Method), J. Hydraul. Res., 7(2), 205–230, 1969.

Dankers, R., Arnell, N. W., Clark, D. B., Falloon, P. D., Fekete, B. M., Gosling, S. N., Heinke, J., Kim, H., Masaki, Y., Satoh, Y., Stacke, T., Wada, Y. and Wisser, D.: First look at changes in flood hazard in the Inter-Sectoral Impact Model Intercomparison Project ensemble, Proc. Natl. Acad. Sci., 111(9), 3257–3261, doi:10.1073/pnas.1302078110, 2014.

Dawson, C. W. and Wilby, R. L.: Hydrological modelling using artificial neural networks, Prog. Phys. Geogr., 25(1), 80–108, doi:10.1177/030913330102500104, 2001.

De Lannoy, G. J. M., Reichle, R. H., Houser, P. R., Pauwels, V. R. N. and Verhoest, N. E. C.: Correcting for forecast bias in soil moisture assimilation with the ensemble Kalman filter, Water Resour. Res., 43(9), W09410, doi:10.1029/2006WR005449, 2007.

Deb, K., Pratap, A., Agarwal, S. and Meyarivan, T.: A fast and elitist multiobjective genetic algorithm: NSGA-II, IEEE Trans. Evol. Comput., 6(2), 182–197, 2002.

Dee, D. P.: On-line Estimation of Error Covariance Parameters for Atmospheric Data Assimilation, Mon. Weather Rev., 123(4), 1128–1145, doi:10.1175/1520-0493(1995)123<1128:OLEOEC>2.0.CO;2, 1995.

Degrossi, L. C., Do Amaral, G. G., da Vasconcelos, E. S. M., Albuquerque, J. P. and Ueyama, J.: Using Wireless Sensor Networks in the Sensor Web for Flood Monitoring in Brazil, in Proceedings of the 10th International ISCRAM Conference, Baden-Baden, Germany, 2013.

Derber, J. and Rosati, A.: A Global Oceanic Data Assimilation System, J. Phys. Oceanogr., 19(9), 1333–1347, doi:10.1175/1520-0485(1989)019<1333:AGODAS>2.0.CO;2, 1989.

DHI: MIKE FLOOD User Manual, 2005.

Di Baldassarre, G. and Montanari, A.: Uncertainty in river discharge observations: a quantitative analysis, Hydrol. Earth Syst. Sci., 13, 913-921, 2009.

Di Baldassarre, G., Castellarin, A. and Brath, A.: Analysis of the effects of levee heightening on flood propagation: example of the River Po, Italy, Hydrol. Sci. J., 54(6), 1007–1017, doi:10.1623/hysj.54.6.1007, 2009.

Di Baldassarre, G., Montanari, A., Lins, H., Koutsoyiannis, D., Brandimarte, L. and Blöschl, G.: Flood fatalities in Africa: From diagnosis to mitigation, Geophys. Res. Lett., 37(22), L22402, doi:10.1029/2010GL045467, 2010.

Di Baldassarre, G., Viglione, A., Carr, G., Kuil, L., Yan, K., Brandimarte, L. and Blöschl, G.: Debates—Perspectives on socio-hydrology: Capturing feedbacks between physical and social processes, Water Resour. Res., 51(6), 4770–4781, doi:10.1002/2014WR016416, 2015.

Dibike, Y. B. and Solomatine, D. P.: River flow forecasting using artificial neural networks, Phys. Chem. Earth Part B Hydrol. Oceans Atmosphere, 26(1), 1–7, doi:10.1016/S1464-1909(01)85005-X, 2001.

Dibike, Y. B., Velickov, S., Solomatine, D. P. and Abbott, M. B.: Model Induction with Support Vector Machines: Introduction and Applications, J. Comput. Civ. Eng., 15(3), 208–216, doi:10.1061/(ASCE)0887-3801(2001)15:3(208), 2001.

Domeneghetti, A., Vorogushyn, S., Castellarin, A., Merz, B. and Brath, A.: Probabilistic flood hazard mapping: effects of uncertain boundary conditions, Hydrol. Earth Syst. Sci., 17(8), 3127–3140, doi:10.5194/hess-17-3127-2013, 2013.

Drecourt, J.-P.: Data assimilation in hydrological modelling, Environment & Resources DTU. Technical University of Denmark., 2004.

Duan, Q., Ajami, N. K., Gao, X. and Sorooshian, S.: Multi-model ensemble hydrologic prediction using Bayesian model averaging, Adv. Water Resour., 30(5), 1371–1386, doi:10.1016/j.advwatres.2006.11.014, 2007.

Duan, Q. Y., Sorooshian, S. and Gupta, V.: Effective and Efficient Global Optimization for Conceptual Rainfall-Runoff Models, Water Resour. Res., 28(4), 1015–1031, doi:10.1029/91wr02985, 1992.

Eckhardt, K.: How to construct recursive digital filters for baseflow separation, Hydrol. Process., 19(2), 507–515, doi:10.1002/hyp.5675, 2005.

Errico, R. M. and Privé, N. C.: An estimate of some analysis-error statistics using the Global Modeling and Assimilation Office observing-system simulation framework, Q. J. R. Meteorol. Soc., 140(680), 1005–1012, doi:10.1002/qj.2180, 2014.

Errico, R. M., Yang, R., Privé, N. C., Tai, K.-S., Todling, R., Sienkiewicz, M. E. and Guo, J.: Development and validation of observing-system simulation experiments at NASA's Global Modeling and Assimilation Office, Q. J. R. Meteorol. Soc., 139(674), 1162–1178, doi:10.1002/qj.2027, 2013.

European Environment Agency: Vulnerability and adaptation to climate change in Europe, Publications Office, Luxembourg. [online] Available from: http://bookshop.europa.eu/uri?target=EUB:NOTICE:THAK06002:EN:HTML (Accessed 19 February 2016), 2006.

Evensen, G.: The Ensemble Kalman Filter: Theoretical formulation and practical implementation, Ocean Dyn., 53(4), 343–367, doi:10.1007/s10236-003-0036-9, 2003.

Evensen, G.: Data Assimilation: The Ensemble Kalman Filter, 2nd ed. 2009 edition., Springer, 2006.

Ferri, M., Monego, M., Norbiato, D., Baruffi, F., Toffolon, C. and Casarin, R.: La piattaforma previsionale per i bacini idrografici del Nord Est Adriatico (I), in Proc.XXXIII Conference of Hydraulics and Hydraulic Engineering, p. 10, Brescia., 2012.

Feyen, L., Vrugt, J. A., Nualláin, B. Ó., van der Knijff, J. and De Roo, A.: Parameter optimisation and uncertainty assessment for large-scale streamflow simulation with the LISFLOOD model, J. Hydrol., 332(3–4), 276–289, doi:10.1016/j.jhydrol.2006.07.004, 2007.

Fischer, C., Montmerle, T., Berre, L., Auger, L. and Ştefănescu, S. E.: An overview of the variational assimilation in the ALADIN/France numerical weather-prediction system, Q. J. R. Meteorol. Soc., 131(613), 3477–3492, doi:10.1256/qj.05.115, 2005.

Fong, L. W.: Multi-sensor track fusion via Multiple-Model Adaptive Filter, in Proceedings of the 48th IEEE Conference on Decision and Control, 2009 held jointly with the 2009 28th Chinese Control Conference. CDC/CCC 2009, pp. 2327–2332., 2009.

Fowler, A. and Jan Van Leeuwen, P.: Observation impact in data assimilation: the effect of non-Gaussian observation error, Tellus A, 65(0), doi:10.3402/tellusa.v65i0.20035, 2013.

García-Pintado, J., Neal, J. C., Mason, D. C., Dance, S. L. and Bates, P. D.: Scheduling satellite-based SAR acquisition for sequential assimilation of water level observations into flood modelling, J. Hydrol., 495, 252–266, doi:10.1016/j.jhydrol.2013.03.050, 2013.

Georgakakos, A. P., Georgakakos, K. P. and Baltas, E. A.: A state-space model for hydrologic river routing, Water Resour. Res., 26(5), 827–838, doi:10.1029/WR026i005p00827, 1990.

Gharesifard, M. and Wehn, U.: To share or not to share: Drivers and barriers for sharing data via online amateur weather networks, J. Hydrol., 535, 181–190, doi:10.1016/j.jhydrol.2016.01.036, 2016.

Giandotti, M.: Previsione delle piene e delle magre dei corsi d'acqua, Servizio Idrografico Italiano, Rome., 1933.

Giustarini, L., Matgen, P., Hostache, R., Montanari, M., Plaza, D., Pauwels, V. R. N., De Lannoy, G. J. M., De Keyser, R., Pfister, L., Hoffmann, L. and Savenije, H. H. G.: Assimilating SAR-derived water level data into a hydraulic model: a case study, Hydrol. Earth Syst. Sci., 15(7), 2349–2365, doi:10.5194/hess-15-2349-2011, 2011.

Götzinger, J. and Bárdossy, A.: Generic error model for calibration and uncertainty estimation of hydrological models, Water Resour. Res., 44(12), W00B07, doi:10.1029/2007WR006691, 2008.

Govindaraju, R. S. and Rao, A. R.: Artificial Neural Networks in Hydrology, Springer Science & Business Media., 2013.

Gura, T.: Citizen science: Amateur experts, Nature, 496(7444), 259–261, doi:10.1038/nj7444-259a, 2013.

Hall, J. and Solomatine, D.: A framework for uncertainty analysis in flood risk management decisions, Int. J. River Basin Manag., 6(2), 85–98, doi:10.1080/15715124.2008.9635339, 2008.

Hannah, D. M., Demuth, S., van Lanen, H. A. J., Looser, U., Prudhomme, C., Rees, G., Stahl, K. and Tallaksen, L. M.: Large-scale river flow archives: importance, current status and future needs, Hydrol. Process., 25(7), 1191–1200, doi:10.1002/hyp.7794, 2011.

Hargreaves, G.H. and Samani, Z.A.: Estimating potential evapotranspiration, J. Irrig. Drain. Eng., 108(3), 223–230, 1982.

Havnø, K., Madsen, M. N. and Dørge, J.: MIKE 11 - a generalized river modelling package., 733–782, 1995.

Heemink, A. W. and Kloosterhuis, H.: Data assimilation for non-linear tidal models, Int. J. Numer. Methods Fluids, 11(8), 1097–1112, doi:10.1002/fld.1650110804, 1990.

Heemink, A. W. and Segers, A. J.: Modeling and prediction of environmental data in space and time using Kalman filtering, Stoch. Environ. Res. Risk Assess., 16(3), 225–240, doi:10.1007/s00477-002-0097-1, 2002.

Heemink, A. W., Verlaan, M. and Segers, A. J.: Variance Reduced Ensemble Kalman Filtering, Mon. Weather Rev., 129(7), 1718–1728, 2001.

Heitmuller, F. T.: Channel adjustments to historical disturbances along the lower Brazos and Sabine Rivers, south-central USA, Geomorphology, 204, 382–398, 2014.

Hinkel, J., Lincke, D., Vafeidis, A. T., Perrette, M., Nicholls, R. J., Tol, R. S. J., Marzeion, B., Fettweis, X., Ionescu, C. and Levermann, A.: Coastal flood damage and adaptation costs under 21st century sea-level rise, Proc. Natl. Acad. Sci., 111(9), 3292–3297, doi:10.1073/pnas.1222469111, 2014.

Horritt, M. S.: A methodology for the validation of uncertain flood inundation models, J. Hydrol., 326(1–4), 153–165, doi:10.1016/j.jhydrol.2005.10.027, 2006.

Hostache, R., Lai, X., Monnier, J. and Puech, C.: Assimilation of spatially distributed water levels into a shallow-water flood model. Part II: Use of a remote sensing image of Mosel River, J. Hydrol., 390(3–4), 257–268, doi:10.1016/j.jhydrol.2010.07.003, 2010.

Houser, P. R., Shuttleworth, W. J., Famiglietti, J. S., Gupta, H. V., Syed, K. H. and Goodrich, D. C.: Integration of soil moisture remote sensing and hydrologic modeling using data assimilation, Water Resour. Res., 34(12), 3405–3420, doi:10.1029/1998WR900001, 1998.

Hover, F. S.: Path planning for data assimilation in mobile environmental monitoring systems, in IEEE/RSJ International Conference on Intelligent Robots and Systems, 2009. IROS 2009, pp. 213–218., 2009.

Howe, J.: Crowdsourcing: Why the Power of the Crowd Is Driving the Future of Business, 1st ed., Crown Publishing Group, New York, NY, USA., 2008.

Huang, B., Kinter, J. L. and Schopf, P. S.: Ocean data assimilation using intermittent analyses and continuous model error correction, Adv. Atmospheric Sci., 19(6), 965–992, doi:10.1007/s00376-002-0059-z, 2002.

Huard, D. and Mailhot, A.: Calibration of hydrological model GR2M using Bayesian uncertainty analysis, Water Resour. Res., 44(2), W02424, doi:10.1029/2007WR005949, 2008.

Hunt, B. R., Kalnay, E., Kostelich, E. J., Ott, E., Patil, D. J., Sauer, T., Szunyogh, I., Yorke, J. A. and Zimin, A. V.: Four-dimensional Ensemble Kalman Filtering, Tellus A, 56(4), 273–277, doi:10.1111/j.1600-0870.2004.00066.x, 2004.

Huwald, H., Barrenetxea, G., de Jong, S., Ferri, M., Carvalho, R., Lanfranchi, V., McCarthy, S., Glorioso, G., Prior, S., Solà, E., Gil-Roldàn, E., Alfonso, L., Wehn de Montalvo, U., Onencan, A., Solomatine, D. and Lobbrecht, A.: D1.11 Sensor technology requirement analysis, Confidential Deliverable, The WeSenseIt Project (FP7/2007-2013 grant agreement no 308429)., 2013.

Ide, K., Courtier, P., Ghil, M. and Lorenc, A. C.: Unifed notation for data assimilation: operational, sequential and variational, J. Meteorol. Soc. Jpn., 75(1B), 181–189, 1997.

ISPUW: iSPUW: Integrated Sensing and Prediction of Urban Water for Sustainable Cities, [online] Available from: http://ispuw.uta.edu/nsf/ (Accessed 19 February 2016), 2015.

Jamieson, D. G., Wilkinson, J. C. and Ibbitt, R. P.: Hydrologic forecasting with sequential deterministic and stochastic stages, Proc. Int. Symp. Uncertainties Hydrol. Water Resour. Syst., 1973.

Jaun, S. and Ahrens, B.: Evaluation of a probabilistic hydrometeorological forecast system, Hydrol. Earth Syst. Sci., 13(7), 1031–1043, 2009.

Ji, S., Hartov, A., Roberts, D. and Paulsen, K.: Data assimilation using a gradient descent method for estimation of intraoperative brain deformation, Med. Image Anal., 13(5), 744–756, doi:10.1016/j.media.2009.07.002, 2009.

Jongman, B., Hochrainer-Stigler, S., Feyen, L., Aerts, J. C. J. H., Mechler, R., Botzen, W. J. W., Bouwer, L. M., Pflug, G., Rojas, R. and Ward, P. J.: Increasing stress on disaster-risk finance due to large floods, Nat. Clim. Change, 4(4), 264–268, doi:10.1038/nclimate2124, 2014.

Journel, A. G. and Huijbregts, C. J.: Mining Geostatistics, Academic Press, London., 1978.

Kachroo, R. K.: River Flow Forecasting River flow forecasting. Part 1. A discussion of the principles, J. Hydrol., 133(1), 1–15, doi:10.1016/0022-1694(92)90146-M, 1992.

Kalman, R. E.: A new approach to linear filtering and prediction problems, J. Basic Eng., 82(1), 35–45, doi:10.1115/1.3662552, 1960.

Kavetski, D., Kuczera, G. and Franks, S. W.: Bayesian analysis of input uncertainty in hydrological modeling: 1. Theory, Water Resour. Res., 42(3), W03407, doi:10.1029/2005WR004368, 2006.

Kim, K. J., Moskowitz, H. and Koksalan, M.: Fuzzy versus statistical linear regression, Eur. J. Oper. Res., 92(2), 417–434, doi:10.1016/0377-2217(94)00352-1, 1996.

Kim, Y., Tachikawa, Y., Shiiba, M., Kim, S., Yorozu, K. and Noh, S. j.: Simultaneous estimation of inflow and channel roughness using 2D hydraulic model and particle filters, J. Flood Risk Manag., 6(2), 112–123, doi:10.1111/j.1753-318X.2012.01164.x, 2013.

Kitanidis, P. K. and Bras, R. L.: Real-time forecasting with a conceptual hydrologic model: 1. Analysis of uncertainty, Water Resour. Res., 16(6), 1025–1033, doi:10.1029/WR016i006p01025, 1980.

Kollat, J. B., Reed, P. M. and Maxwell, R. M.: Many-objective groundwater monitoring network design using bias-aware ensemble Kalman filtering, evolutionary optimization, and visual analytics, Water Resour. Res., 47(2), W02529, doi:10.1029/2010WR009194, 2011.

Komma, J., Blöschl, G. and Reszler, C.: Soil moisture updating by Ensemble Kalman Filtering in real-time flood forecasting, J. Hydrol., 357(3–4), 228–242, doi:10.1016/j.jhydrol.2008.05.020, 2008.

Koussis, A.D.: Discussion of "Accuracy Criteria in Diffusion Routing" by Victor Miguel Ponce and Fred D. Theurer (June, 1982), J. Hydraul. Eng., 109(5), 803–806, doi:10.1061/(ASCE)0733-9429(1983)109:5(803), 1983.

Koutsoyiannis, D.: HESS Opinions "A random walk on water," Hydrol Earth Syst Sci, 14(3), 585–601, doi:10.5194/hess-14-585-2010, 2010.

Kovitz, J. L. and Christakos, G.: Assimilation of fuzzy data by the BME method, Stoch. Environ. Res. Risk Assess., 18(2), 79–90, doi:10.1007/s00477-003-0128-6, 2004.

Krstanovic, P. F. and Singh, V. P.: Evaluation of rainfall networks using entropy: II. Application, Water Resour. Manag., 6(4), 295–314, doi:10.1007/BF00872282, 1992.

Krzysztofowicz, R.: Bayesian theory of probabilistic forecasting via deterministic hydrologic model, Water Resour. Res., 35(9), 2739–2750, doi:10.1029/1999WR900099, 1999.

Krzysztofowicz, R.: The case for probabilistic forecasting in hydrology, J. Hydrol., 249(1–4), 2–9, doi:10.1016/S0022-1694(01)00420-6, 2001.

Kuczera, G. and Parent, E.: Monte Carlo assessment of parameter uncertainty in conceptual catchment models: the Metropolis algorithm, J. Hydrol., 211(1–4), 69–85, doi:10.1016/S0022-1694(98)00198-X, 1998.

Kuczera, G., Kavetski, D., Franks, S. and Thyer, M.: Towards a Bayesian total error analysis of conceptual rainfall-runoff models: Characterising model error using storm-dependent parameters, J. Hydrol., 331(1–2), 161–177, doi:10.1016/j.jhydrol.2006.05.010, 2006.

Kumar, R., Chatterjee, C., Lohani, A. K., Kumar, S. and Singh, R. D.: Sensitivity Analysis of the GIUH based Clark Model for a Catchment, Water Resour. Manag., 16(4), 263–278, doi:10.1023/A:1021920717410, 2002.

Lahoz, W., Khattatov, B. and Menard, R.: Data Assimilation: Making Sense of Observations, Springer Science & Business Media., 2010.

Lahoz, W. A. and Schneider, P.: Data assimilation: making sense of Earth Observation, Front. Environ. Sci., 2, doi:10.3389/fenvs.2014.00016, 2014.

Laio, F., Porporato, A., Ridolfi, L. and Rodriguez-Iturbe, I.: Plants in water-controlled ecosystems: active role in hydrologic processes and response to water stress: II. Probabilistic soil moisture dynamics, Adv. Water Resour., 24(7), 707–723, doi:10.1016/S0309-1708(01)00005-7, 2001.

Lee, H., Seo, D. J. and Koren, V.: Assimilation of streamflow and in situ soil moisture data into operational distributed hydrologic models: Effects of uncertainties in the data and initial model soil moisture states, Adv. Water Resour., 34(12), 1597–1615, doi:10.1016/j.advwatres.2011.08.012, 2011a.

Lee, H., Liu, Y., He, M., Demargne, J. and Seo, D. J.: Variational assimilation of streamflow into three-parameter Muskingum routing model for improved operational river flow forecasting, 13, EGU2011-13073, EGU General Assembly, Vienna. 2011.

Lee, H., Seo, D. J., Liu, Y., Koren, V., McKee, P. and Corby, R.: Variational assimilation of streamflow into operational distributed hydrologic models: Effect of spatiotemporal scale of adjustment, Hydrol. Earth Syst. Sci., 16(7), 2233–2251, doi:10.5194/hess-16-2233-2012, 2012.

Li, Z. and Navon, I. M.: Optimality of variational data assimilation and its relationship with the Kalman filter and smoother, Q. J. R. Meteorol. Soc., 127(572), 661–683, doi:10.1002/qj.49712757220, 2001.

Lindström, G., Johansson, B., Persson, M., Gardelin, M. and Bergström, S.: Development and test of the distributed HBV-96 hydrological model, J. Hydrol., 201(1–4), 272–288, doi:10.1016/S0022-1694(97)00041-3, 1997.

Liu, Y. and Gupta, H. V.: Uncertainty in hydrologic modeling: Toward an integrated data assimilation framework, Water Resour. Res., 43(7), 1–18, doi:10.1029/2006WR005756, 2007.

Liu, Y., Lee, H., Seo, D., Brown, J., Corby, R. and Howieson, T.: Ensemble Data Assimilation for Channel Flow Routing to Improve Operational Hydrologic Forecasting, AGU Fall Meet. Abstr., 51 [online] Available from: http://adsabs.harvard.edu/abs/2008AGUFM.H51E0868L, 2008.

Liu, Y., Weerts, A. H., Clark, M., Hendricks Franssen, H. J., Kumar, S., Moradkhani, H., Seo, D. J., Schwanenberg, D., Smith, P., Van Dijk, A. I. J. M., Van Velzen, N., He, M., Lee, H., Noh, S. J., Rakovec, O. and Restrepo, P.: Advancing data assimilation in operational hydrologic forecasting: Progresses, challenges, and emerging opportunities, Hydrol. Earth Syst. Sci., 16(10), 3863–3887, doi:10.5194/hess-16-3863-2012, 2012.

Liu, Z., Martina, M. L. and Todini, E.: Flood forecasting using a fully distributed model: application of the TOPKAPI model to the Upper Xixian Catchment, Hydrol. Earth Syst. Sci. Discuss., 9(4), 347–364, 2005.

Lopez Lopez, P., Wanders, N., Schellekens, J., Renzullo, L. J., Sutanudjaja, E. H. and Bierkens, M. F. P.: Improved large-scale hydrological modelling through the assimilation of streamflow and downscaled satellite soil moisture observations, Hydrol. Earth Syst. Sci. Discuss., 12(10), 10559–10601, doi:10.5194/hessd-12-10559-2015, 2015.

Lorenc, A. C. and Rawlins, F.: Why does 4D-Var beat 3D-Var?, Q. J. R. Meteorol. Soc., 131(613), 3247–3257, doi:10.1256/qj.05.85, 2005.

Lowry, C. S. and Fienen, M. N.: CrowdHydrology: Crowdsourcing hydrologic data and engaging citizen scientists, GroundWater, 51(1), 151–156, doi:10.1111/j.1745-6584.2012.00956.x, 2013.

Lü, H., Yu, Z., Zhu, Y., Drake, S., Hao, Z. and Sudicky, E. A.: Dual state-parameter estimation of root zone soil moisture by optimal parameter estimation and extended Kalman filter data assimilation, Adv. Water Resour., 34(3), 395–406, doi:10.1016/j.advwatres.2010.12.005, 2011.

Lü, H., Hou, T., Horton, R., Zhu, Y., Chen, X., Jia, Y., Wang, W. and Fu, X.: The streamflow estimation using the Xinanjiang rainfall runoff model and dual state-parameter estimation method, J. Hydrol., 480, 102–114, doi:10.1016/j.jhydrol.2012.12.011, 2013.

Lundberg, A.: Combination of a conceptual model and an autoregressive error model for improving short time forecasting, Hydrol. Res., 13(4), 233–246, 1982.

Lunn, K. E., Paulsen, K. D., Liu, F., Kennedy, F. E., Hartov, A. and Roberts, D. W.: Data-guided brain deformation modeling: evaluation of a 3-D adjoint inversion method in porcine studies, IEEE Trans. Biomed. Eng., 53(10), 1893–1900, doi:10.1109/TBME.2006.881771, 2006.

Macpherson, B.: Dynamic initialization by repeated insertion of data, Q. J. R. Meteorol. Soc., 117(501), 965–991, doi:10.1002/qj.49711750105, 1991.

Madsen, H. and Cañizares, R.: Comparison of extended and ensemble Kalman filters for data assimilation in coastal area modelling, Int. J. Numer. Methods Fluids, 31(6), 961–981, doi:10.1002/(SICI)1097-0363(19991130)31:6<961::AID-FLD907>3.0.CO;2-0, 1999.

Madsen, H. and Skotner, C.: Adaptive state updating in real-time river flow forecasting - A combined filtering and error forecasting procedure, J. Hydrol., 308(1–4), 302–312, doi:10.1016/j.jhydrol.2004.10.030, 2005.

Madsen, H., Rosbjerg, D., Damgard, J. and Hansen, F. S.: Data assimilation in the MIKE 11 Flood Forecasting system using Kalman filtering, Int. Assoc. Hydrol. Sci. Publ., 281, 75–81, 2003.

Mantovan, P. and Todini, E.: Hydrological forecasting uncertainty assessment: Incoherence of the GLUE methodology, J. Hydrol., 330(1–2), 368–381, doi:10.1016/j.jhydrol.2006.04.046, 2006.

Marshall, L., Nott, D. and Sharma, A.: Towards dynamic catchment modelling: a Bayesian hierarchical mixtures of experts framework, Hydrol. Process., 21(7), 847–861, doi:10.1002/hyp.6294, 2007.

Mason, D. C., Schumann, G. J.-P., Neal, J. C., Garcia-Pintado, J. and Bates, P. D.: Automatic near real-time selection of flood water levels from high resolution Synthetic Aperture Radar images for assimilation into hydraulic models: A case study, Remote Sens. Environ., 124, 705–716, doi:10.1016/j.rse.2012.06.017, 2012.

Massart, S., Pajot, B., Piacentini, A. and Pannekoucke, O.: On the merits of using a 3D-FGAT assimilation scheme with an outer loop for atmospheric situations governed by transport, Mon. Weather Rev., 138(12), 4509–4522, 2010.

Matgen, P., Montanari, M., Hostache, R., Pfister, L., Hoffmann, L., Plaza, D., Pauwels, V. R. N., De Lannoy, G. J. M., De Keyser, R. and Savenije, H. H. G.: Towards the sequential assimilation of SAR-derived water stages into hydraulic models using the Particle Filter: Proof of concept, Hydrol. Earth Syst. Sci., 14(9), 1773–1785, doi:10.5194/hess-14-1773-2010, 2010.

Matheron, G.: Principles of geostatistics, Econ. Geol., 58(8), 1246–1266, 1963.

Maybeck, P. S.: Stochastic Models, Estimation, and Control, Academic Press., 1982.

Mazzoleni, M., Bacchi, B., Barontini, S., Di Baldassarre, G., Pilotti, M. and Ranzi, R.: Flooding Hazard Mapping in Floodplain Areas Affected by Piping Breaches in the Po River, Italy, J. Hydrol. Eng., 19(4), 717–731, doi:10.1061/(ASCE)HE.1943-5584.0000840, 2014.

Mazzoleni, M., Alfonso, L., Chacon-Hurtado, J. and Solomatine, D.: Assimilating uncertain, dynamic and intermittent streamflow observations in hydrological models, Adv. Water Resour., 83, 323–339, 2015a.

Mazzoleni, M., Verlaan, M., Alfonso, L., Monego, M., Norbiato, D., Ferri, M. and Solomatine, D. P.: Can assimilation of crowdsourced streamflow observations in hydrological modelling improve flood prediction?, Hydrol. Earth Syst. Sci. Discuss., 12(11), 11371–11419, 2015b.

Mazzoleni, M., Barontini, S., Ranzi, R. and Brandimarte, L.: Innovative Probabilistic Methodology for Evaluating the Reliability of Discrete Levee Reaches Owing to Piping, J. Hydrol. Eng., 20(5), 4014067, doi:10.1061/(ASCE)HE.1943-5584.0001055, 2015c.

Mazzoleni, M., Alfonso, L. and Solomatine, D. P.: Influence of spatial distribution of sensors and observation accuracy on the assimilation of distributed streamflow data in hydrological modeling, Hydrol. Sci. J., accepted, 2016.

McCabe, M. F., Wood, E. F., Wójcik, R., Pan, M., Sheffield, J., Gao, H. and Su, H.: Hydrological consistency using multi-sensor remote sensing data for water and energy cycle studies, Remote Sens. Environ., 112(2), 430–444, doi:10.1016/j.rse.2007.03.027, 2008.

McDaniel, J., Kostelich, E., Kuang, Y., Nagy, J., Preul, M. C., Moore, N. Z. and Matirosyan, N. L.: Data assimilation in brain tumor models, in Mathematical

Methods and Models in Biomedicine, pp. 233–262, Springer. [online] Available from: http://link.springer.com/chapter/10.1007/978-1-4614-4178-6_9, 2013.

McDonnell, J. J. and Beven, K.: Debates—The future of hydrological sciences: A (common) path forward? A call to action aimed at understanding velocities, celerities and residence time distributions of the headwater hydrograph, Water Resour. Res., 50(6), 5342–5350, doi:10.1002/2013WR015141, 2014.

McLaughlin, D.: Recent developments in hydrologic data assimilation, Rev. Geophys., 33(95), 977–984, 1995.

McLaughlin, D.: An integrated approach to hydrologic data assimilation: Interpolation, smoothing, and filtering, Adv. Water Resour., 25(8–12), 1275–1286, doi:10.1016/S0309-1708(02)00055-6, 2002.

McMillan, H. K., Jackson, B., Clark, M., Kavetski, D. and Woods, R.: Rainfall uncertainty in hydrological modelling: An evaluation of multiplicative error models, J. Hydrol., 400(1–2), 83–94, doi:10.1016/j.jhydrol.2011.01.026, 2011.

McMillan, H. K., Hreinsson, E. O., Clark, M. P., Singh, S. K., Zammit, C. and Uddstrom, M. J.: Operational hydrological data assimilation with the recursive ensemble Kalman filter, Hydrol. Earth Syst. Sci., 17(1), 21–38, doi:10.5194/hess-17-21-2013, 2013.

Mendoza, P. A., McPhee, J. and Vargas, X.: Uncertainty in flood forecasting: A distributed modeling approach in a sparse data catchment, Water Resour. Res., 48(9), doi:10.1029/2011WR011089, 2012.

Merz, B. and Thieken, A. H.: Separating natural and epistemic uncertainty in flood frequency analysis, J. Hydrol., 309(1–4), 114–132, doi:10.1016/j.jhydrol.2004.11.015, 2005.

Moel, H. de and Aerts, J. C. J. H.: Effect of uncertainty in land use, damage models and inundation depth on flood damage estimates, Nat. Hazards, 58(1), 407–425, doi:10.1007/s11069-010-9675-6, 2010.

Montaldo, N., Albertson, J. D. and Mancini, M.: Dynamic Calibration with an Ensemble Kalman Filter Based Data Assimilation Approach for Root-Zone Moisture Predictions, J. Hydrometeorol., 8(4), 910–921, doi:10.1175/JHM582.1, 2007.

Montanari, A. and Koutsoyiannis, D.: A blueprint for process-based modeling of uncertain hydrological systems, Water Resour. Res., 48(9), W09555, doi:10.1029/2011WR011412, 2012.

Montanari, M., Hostache, R., Matgen, P., Schumann, G., Pfister, L. and Hoffmann, L.: Calibration and sequential updating of a coupled hydrologic-hydraulic model using remote sensing-derived water stages, Hydrol. Earth Syst. Sci., 5(6), 3213–3245, doi:10.5194/hessd-5-3213-2008, 2008.

Montzka, C., Pauwels, V., Franssen, H.-J., Han, X. and Vereecken, H.: Multivariate and Multiscale Data Assimilation in Terrestrial Systems: A Review, Sensors, 12(12), 16291–16333, doi:10.3390/s121216291, 2012.

Moore, R. J.: The probability-distributed principle and runoff production at point and basin scales, Hydrol. Sci. J., 30(2), 273–297, doi:10.1080/02626668509490989, 1985.

Moore, R. J.: The PDM rainfall-runoff model, Hydrol. Earth Syst. Sci. Discuss., 11(1), 483–499, 2007.

Moore, R. J., Jones, D. A., Cox, D. R. and Isham, V. S.: Design of the HYREX raingauge network, Hydrol. Earth Syst. Sci., 4(4), 521–530, doi:10.5194/hess-4-521-2000, 2000.

Moradkhani, H., Sorooshian, S., Gupta, H. V. and Houser, P. R.: Dual state-parameter estimation of hydrological models using ensemble Kalman filter, Adv. Water Resour., 28(2), 135–147, doi:10.1016/j.advwatres.2004.09.002, 2005a.

Moradkhani, H., Hsu, K.-L., Gupta, H. and Sorooshian, S.: Uncertainty assessment of hydrologic model states and parameters: Sequential data assimilation using the particle filter, Water Resour. Res., 41(5), W05012, doi:10.1029/2004WR003604, 2005b.

Moulin, L., Gaume, E. and Obled, C.: Uncertainties on mean areal precipitation: assessment and impact on streamflow simulations, Hydrol Earth Syst Sci, 13(2), 99–114, doi:10.5194/hess-13-99-2009, 2009.

Moulton, K. M., Cornell, A. and Petriu, E.: A fuzzy error correction control system, IEEE Trans. Instrum. Meas., 50(5), 1456–1463, doi:10.1109/19.963224, 2001.

Mukolwe, M. M., Yan, K., Baldassarre, G. D. and Solomatine, D. P.: Testing new sources of topographic data for flood propagation modelling under structural, parameter and observation uncertainty, Hydrol. Sci. J., doi:10.1080/02626667.2015.1019507, 2015.

Murphy, J. M.: The impact of ensemble forecasts on predictability, Q. J. R. Meteorol. Soc., 114(480), 463–493, doi:10.1002/qj.49711448010, 1988.

Nash, J. E. and Sutcliffe, J. V.: River flow forecasting through conceptual models part I — A discussion of principles, J. Hydrol., 10(3), 282–290, doi:10.1016/0022-1694(70)90255-6, 1970.

Neal, J., Schumann, G., Bates, P., Buytaert, W., Matgen, P. and Pappenberger, F.: A data assimilation approach to discharge estimation from space, Hydrol. Process., 23(25), 3641–3649, doi:10.1002/hyp.7518, 2009.

Neal, J. C., Atkinson, P. M. and Hutton, C. W.: Flood inundation model updating using an ensemble Kalman filter and spatially distributed measurements, J. Hydrol., 336(3–4), 401–415, doi:10.1016/j.jhydrol.2007.01.012, 2007.

Neal, J. C., Atkinson, P. M. and Hutton, C. W.: Adaptive space–time sampling with wireless sensor nodes for flood forecasting, J. Hydrol., 414–415, 136–147, doi:10.1016/j.jhydrol.2011.10.021, 2012.

Nester, T., Komma, J., Viglione, A. and Blöschl, G.: Flood forecast errors and ensemble spread—A case study, Water Resour. Res., 48(10), W10502, doi:10.1029/2011WR011649, 2012.

Noh, S., Tachikawa, Y., Shiiba, M. and Kim, S.: Ensemble Kalman Filtering and Particle Filtering in a Lag-Time Window for Short-Term Streamflow Forecasting with a Distributed Hydrologic Model, J. Hydrol. Eng., 18(12), 1684–1696, doi:10.1061/(ASCE)HE.1943-5584.0000751, 2013.

Noh, S. J., Rakovec, O., Weerts, A. H. and Tachikawa, Y.: On noise specification in data assimilation schemes for improved flood forecasting using distributed hydrological models, J. Hydrol., 519, Part D, 2707–2721, doi:10.1016/j.jhydrol.2014.07.049, 2014.

O'Donnell, T.: A direct three-parameter Muskingum procedure incorporating lateral inflow, Hydrol. Sci. J., 30(4), 479–496, doi:10.1080/02626668509491013, 1985.

Özelkan, E. C. and Duckstein, L.: Multi-objective fuzzy regression: a general framework, Comput. Oper. Res., 27(7–8), 635–652, doi:10.1016/S0305-0548(99)00110-0, 2000.

Pan, M., Wood, E. F., Wójcik, R. and McCabe, M. F.: Estimation of regional terrestrial water cycle using multi-sensor remote sensing observations and data assimilation, Remote Sens. Environ., 112(4), 1282–1294, doi:10.1016/j.rse.2007.02.039, 2008.

Pappenberger, F. and Beven, K. J.: Ignorance is bliss: Or seven reasons not to use uncertainty analysis, Water Resour. Res., 42(5), 1–8, doi:10.1029/2005WR004820, 2006.

Pappenberger, F., Matgen, P., Beven, K. J., Henry, J.-B., Pfister, L. and Fraipont de, P.: Influence of uncertain boundary conditions and model structure on flood inundation predictions, Adv. Water Resour., 29(10), 1430–1449, doi:10.1016/j.advwatres.2005.11.012, 2006.

Park, S. K. and Xu, L.: Data Assimilation for Atmospheric, Oceanic and Hydrologic Applications, Springer Science & Business Media., 2013.

Pauwels, V. R. N. and De Lannoy, G. J. M.: Improvement of Modeled Soil Wetness Conditions and Turbulent Fluxes through the Assimilation of Observed Discharge, J. Hydrometeorol., 7(3), 458–477, doi:10.1175/JHM490.1, 2006.

Pauwels, V. R. N. and De Lannoy, G. J. M.: Ensemble-based assimilation of discharge into rainfall-runoff models: A comparison of approaches to mapping observational information to state space, Water Resour. Res., 45(8), W08428, doi:10.1029/2008WR007590, 2009.

Perrin, C., Michel, C. and Andréassian, V.: Improvement of a parsimonious model for streamflow simulation, J. Hydrol., 279(1–4), 275–289, doi:10.1016/S0022-1694(03)00225-7, 2003.

Phillips, J. D.: Geomorphic controls and transition zones in the lower Sabine River, Hydrol. Process., 22(14), 2424–2437, 2008.

Pipunic, R. C., Walker, J. P., Western, A. W. and Trudinger, C. M.: Assimilation of multiple data types for improved heat flux prediction: A one-dimensional field study, Remote Sens. Environ., 136, 315–329, doi:10.1016/j.rse.2013.05.015, 2013.

Plate, E. and Shahzad, K.: Uncertainty Analysis of Multi-Model Flood Forecasts, Water, 7(12), 6788–6809, doi:10.3390/w7126654, 2015.

Ponce, V. M. and Changanti, P. V.: Variable-parameter Muskingum-Cunge method revisited, J. Hydrol., 162(3–4), 433–439, doi:10.1016/0022-1694(94)90241-0, 1994.

Ponce, V. M. and Lugo, A.: Modeling Looped Ratings in Muskingum-Cunge Routing, J. Hydrol. Eng., 6(2), 119–124, doi:10.1061/(ASCE)1084-0699(2001)6:2(119), 2001.

Preissmann, A.: Propagation of Translatory Waves in Channels and Rivers, in in Proc., First Congress of French Assoc. for Computation, pp. 433–442, Grenoble, France., 1961.

Press, W.H., Teukolsky, S.A., Vetterling, W.T. and Flannery, B.P.: Numerical recipes in fortran. 2nd ed., 1992.

Puente, C. E. and Bras, R. L.: Application of nonlinear filtering in the real time forecasting of river flows, Water Resour. Res., 23(4), 675–682, doi:10.1029/WR023i004p00675, 1987.

Quinonero-Candela, J., Rasmussen, C. E., Sinz, F., Bousquet, O. and Schölkopf, B.: Evaluating predictive uncertainty challenge, in Machine Learning Challenges. Evaluating Predictive Uncertainty, Visual Object Classification, and Recognising Tectual Entailment, 1–27, Springer. [online] Available from: http://link.springer.com/10.1007%2F11736790_1 (Accessed 2 March 2016), 2006.

Rabuñal, J. R., Puertas, J., Suárez, J. and Rivero, D.: Determination of the unit hydrograph of a typical urban basin using genetic programming and artificial neural networks, Hydrol. Process., 21(4), 476–485, doi:10.1002/hyp.6250, 2007.

Rafieeinasab, A., Seo, D.-J., Lee, H. and Kim, S.: Comparative evaluation of maximum likelihood ensemble filter and ensemble Kalman filter for real-time assimilation of streamflow data into operational hydrologic models, J. Hydrol., 519, Part D, 2663–2675, doi:10.1016/j.jhydrol.2014.06.052, 2014.

Ragnoli, E., Zhuk, S., Donncha, F. O., Suits, F. and Hartnett, M.: An optimal interpolation scheme for assimilation of HF radar current data into a numerical ocean model, in Oceans, 2012, pp. 1–5., 2012.

Raiffa, H. and Schlaifer, R.: Applied Statistical Decision Theory, MIT Press, Cambridge, MA., 1961.

Rakovec, O., Weerts, A. H., Hazenberg, P., F. Torfs, P. J. J. and Uijlenhoet, R.: State updating of a distributed hydrological model with ensemble kalman Filtering: Effects of updating frequency and observation network density on forecast accuracy, Hydrol. Earth Syst. Sci., 16(9), 3435–3449, doi:10.5194/hess-16-3435-2012, 2012.

Rakovec, O., Weerts, A. H., Sumihar, J. and Uijlenhoet, R.: Operational aspects of asynchronous filtering for flood forecasting, Hydrol. Earth Syst. Sci., 19(6), 2911–2924, doi:10.5194/hess-19-2911-2015, 2015.

Ranzi, R., Mazzoleni, M., Milanesi, L., Pilotti, M., Ferri, M., Giurato, F., Michel, G., Fewtrell, T., Bates, P., Neal, J., Di Baldassarre, G., Bogaard, T., Brilly, M. and Mikos, M.: Critical review of non-structural measures for water-related risks, Project Deliverable., 2011.

Rao, K. H. V. D., Rao, V. V., Dadhwal, V. K., Behera, G. and Sharma, J. R.: A distributed model for real-time flood forecasting in the Godavari Basin using space inputs, Int. J. Disaster Risk Sci., 2(3), 31–40, doi:10.1007/s13753-011-0014-7, 2011.

Rasmussen, J., Madsen, H., Jensen, K. H. and Refsgaard, J. C.: Data assimilation in integrated hydrological modeling using ensemble Kalman filtering: evaluating the effect of ensemble size and localization on filter performance, Hydrol. Earth Syst. Sci., 19(7), 2999–3013, doi:10.5194/hess-19-2999-2015, 2015.

Refsgaard, J. C.: Validation and Intercomparison of Different Updating Procedures for Real-Time Forecasting, Nord. Hydrol., 28(2), 65–84, doi:10.2166/nh.1997.005, 1997.

Reggiani, P., Sivapalan, M. and Hassanizadeh, S. M.: A unifying framework for watershed thermodynamics: balance equations for mass, momentum, energy and entropy, and the second law of thermodynamics, Adv. Water Resour., 22(4), 367–398, 1998.

Reichle, R., McLaughlin, D. B. and Entekhabi, D.: Hydrologic data assimilation with the ensemble Kalman filter, Am. Meteorol. Soc., 130(1), 103–114, doi:10.1175/1520-0493(2002)130<0103:HDAWTE>2.0.CO;2, 2002.

Reichle, R. H.: Data assimilation methods in the Earth sciences, Adv. Water Resour., 31(11), 1411–1418, doi:10.1016/j.advwatres.2008.01.001, 2008.

Reichle, R. H., Crow, W. T. and Keppenne, C. L.: An adaptive ensemble Kalman filter for soil moisture data assimilation, Water Resour. Res., 44(3), W03423, doi:10.1029/2007WR006357, 2008.

Reichle, R. H. : Variational assimilation of remote sensing data for land surface hydrologic applications, Thesis, Massachusetts Institute of Technology. [online] Available from: http://dspace.mit.edu/handle/1721.1/28220, 2000.

Renard, B., Kavetski, D., Kuczera, G., Thyer, M. and Franks, S. W.: Understanding predictive uncertainty in hydrologic modeling: The challenge of identifying input and structural errors, Water Resour. Res., 46(5), W05521, doi:10.1029/2009WR008328, 2010.

Ricci, S., Piacentini, A., Thual, O., Le Pape, E. and Jonville, G.: Correction of upstream flow and hydraulic state with data assimilation in the context of flood forecasting, Hydrol Earth Syst Sci, 15(11), 3555–3575, doi:10.5194/hess-15-3555-2011, 2011.

Ridolfi, E., Alfonso, L., Baldassarre, G. D., Dottori, F., Russo, F. and Napolitano, F.: An entropy approach for the optimization of cross-section spacing for river modelling, Hydrol. Sci. J., 59(1), 126–137, doi:10.1080/02626667.2013.822640, 2014.

Rinaldo, A. and Rodriguez-Iturbe, I.: Geomorphological Theory of the Hydrological Response, Hydrol. Process., 10(6), 803–829, doi:10.1002/(SICI)1099-1085(199606)10:6<803::AID-HYP373>3.0.CO;2-N, 1996.

Robinson, A. R., Lermusiaux, P. F. J. and Sloan III, N. Q.: Data assimilation, The sea, 10, 541–594, 1998.

Rodríguez-Iturbe, I., González-Sanabria, M. and Bras, R. L.: A geomorphoclimatic theory of the instantaneous unit hydrograph, Water Resour. Res., 18(4), 877–886, doi:10.1029/WR018i004p00877, 1982.

Romanowicz, R. J., Young, P. C. and Beven, K. J.: Data assimilation and adaptive forecasting of water levels in the river Severn catchment, United Kingdom, Water Resour. Res., 42(6), 1–12, doi:10.1029/2005WR004373, 2006.

de Roo, A. P., Gouweleeuw, B., Thielen, J., Bartholmes, J., Bongioannini-Cerlini, P., Todini, E., Bates, P. D., Horritt, M., Hunter, N., Beven, K. and others: Development of a European flood forecasting system, Int. J. River Basin Manag., 1(1), 49–59, 2003.

Rotman, D., Preece, J., Hammock, J., Procita, K., Hansen, D., Parr, C., Lewis, D. and Jacobs, D.: Dynamic Changes in Motivation in Collaborative Citizen-Science Projects, Proc. ACM 2012 Conf. Comput. Support. Coop. Work - CSCW 12, 217–226, doi:10.1145/2145204.2145238, 2012.

Roy, H. E., Pocock, M. J. O., Preston, C. D., Roy, D. B. and Savage, J.: Understanding Citizen Science and Environmental Monitoring, Final Report of UK Environmental Observation Framework., 2012.

Royem, A. A., Mui, C. K., Fuka, D. R. and Walter, M. T.: Technical note: proposing a low-tech, affordable, accurate stream stage monitoring system, Trans. ASABE, 55(6), 2237–2242, 2012.

Sakov, P., Evensen, G. and Bertino, L.: Asynchronous data assimilation with the EnKF, Tellus A, 62(1), 24–29, doi:10.1111/j.1600-0870.2009.00417.x, 2010.

Salamon, P. and Feyen, L.: Assessing parameter, precipitation, and predictive uncertainty in a distributed hydrological model using sequential data assimilation with the particle filter, J. Hydrol., 376(3–4), 428–442, doi:10.1016/j.jhydrol.2009.07.051, 2009.

Savelieva, E., Demyanov, V. and Maignan, M.: Geostatistics: Spatial Predictions and Simulations, in Advanced Mapping of Environmental Data: Geostatistics, Machine Learning and Bayesian Maximum Entropy, edited by M. Kanevski, ISTE, London, UK., 2008.

Schneider, P., Castell, N., Vogt, M., Lahoz, W. and Bartonova, A.: Making sense of crowdsourced observations: Data fusion techniques for real-time mapping of urban air quality, vol. 17, p. 3503. [online] Available from: http://adsabs.harvard.edu/abs/2015EGUGA..17.3503S, 2015.

Schumann, G., Bates, P. D., Horritt, M. S., Matgen, P. and Pappenberger, F.: Progress in intergration of remote sensing derived flood extent and stage data and hydraulic models, Rev. Geophys., 47(2008), 1–20, doi:10.1029/2008RG000274.1.INTRODUCTION, 2009.

Seibert, J. and Beven, K. J.: Gauging the ungauged basin : how many discharge measurements are needed?, Hydrol. Earth Syst. Sci., 13, 883–892, doi:10.5194/hessd-6-2275-2009, 2009.

Seibert, J. and McDonnell, J. J.: On the dialog between experimentalist and modeler in catchment hydrology: Use of soft data for multicriteria model calibration, Water Resour. Res., 38(11), 1241, doi:10.1029/2001WR000978, 2002.

Seo, D. ., Kerke, B., Zink, M., Fang, N., Gao, J. and Yu, X.: iSPUW: A Vision for Integrated Sensing and Prediction of Urban Water for Sustainable Cities., 2014.

Seo, D. J., Koren, V. and Cajina, N.: Real-Time Variational Assimilation of Hydrologic and Hydrometeorological Data into Operational Hydrologic Forecasting, J. Hydrometeorol., 4(3), 627–641, doi:10.1175/1525-7541(2003)004<0627:RVAOHA>2.0.CO;2, 2003.

Seo, D. J., Cajina, L., Corby, R. and Howieson, T.: Automatic state updating for operational streamflow forecasting via variational data assimilation, J. Hydrol., 367(3–4), 255–275, doi:10.1016/j.jhydrol.2009.01.019, 2009.

Shamsudin, S. and Hashim, N.: Rainfall runoff simulation using MIKE 11 NAM, J. Civ. Eng., 15(2), 1–13, 2002.

Shanley, L., Burns, R., Bastian, Z. and Robson, E.: Tweeting up a storm: the promise and perils of crisis mapping, Available SSRN 2464599 [online] Available from: http://papers.ssrn.com/sol3/papers.cfm?abstract_id=2464599, 2013.

Sheffield, J., Goteti, G. and Wood, E. F.: Development of a 50-Year High-Resolution Global Dataset of Meteorological Forcings for Land Surface Modeling, J. Clim., 19(13), 3088–3111, doi:10.1175/JCLI3790.1, 2006.

Shima, T., Oshman, Y. and Shinar, J.: Efficient multiple model adaptive estimation in ballistic missile interception scenarios, J. Guid. Control Dyn., 25(4), 667–675, 2002.

Shrestha, D. L. and Solomatine, D. P.: Data-driven approaches for estimating uncertainty in rainfall-runoff modelling, Int. J. River Basin Manag., 6(2), 109–122, doi:10.1080/15715124.2008.9635341, 2008.

Shrestha, D. L., Kayastha, N. and Solomatine, D. P.: A novel approach to parameter uncertainty analysis of hydrological models using neural networks, Hydrol. Earth Syst. Sci., 13(7), 1235–1248, doi:10.5194/hess-13-1235-2009, 2009.

Sinopoli, B., Schenato, L., Franceschetti, M., Poolla, K., Jordan, M. I. and Sastry, S. S.: Kalman filtering with intermittent observations, in 42nd IEEE Conference on Decision and Control, 2003. Proceedings, 1, 701–708, 2003.

Sittner, W. T. and Krouse, K. .: Improvement of hydrologic simulation by utilizing observed discharge as an indirect input:, U.S. Dept. of Commerce, National Oceanic and Atmospheric Administration, National Weather Service., 1979.

Solomatine, D. P. and Wagener, T.: Hydrological Modelling, in Treatise on Water Science (Wilderer, ed.), 435–457, Elsevier., 2011.

Statistica: Smartphone penetration in Italy (share of mobile users), Statista [online] Available from: http://www.statista.com/statistics/257053/smartphone-user-penetration-in-italy/ (Accessed 20 March 2016), 2016.

Stauffer, D. R. and Seaman, N. L.: Use of Four-Dimensional Data Assimilation in a Limited-Area Mesoscale Model. Part I: Experiments with Synoptic-Scale Data, Mon. Weather Rev., 118(6), 1250–1277, doi:10.1175/1520-0493(1990)118<1250:UOFDDA>2.0.CO;2, 1990.

Stedinger, J. R., Vogel, R. M., Lee, S. U. and Batchelder, R.: Appraisal of the generalized likelihood uncertainty estimation (GLUE) method, Water Resour. Res., 44(12), W00B06, doi:10.1029/2008WR006822, 2008.

Strahler, A. N.: Quantitative analysis of watershed geomorphology, Trans. Am. Geophys. Union, 38(6), 913–913, doi:10.1029/TR038i006p00913, 1957.

Sun, L., Seidou, O., Nistor, I. and Liu, K.: Review of the Kalman type hydrological data assimilation, Hydrol. Sci. J., 0(ja), null, doi:10.1080/02626667.2015.1127376, 2015.

Szilagyi, J. and Szollosi-Nagy, A.: Recursive Streamflow Forecasting: A State Space Approach - CRC Press Book., 2010.

Teisberg, T. J. and Weiher, R. F.: Background paper on the benefits and costs of early warning systems for major natural hazards, World Bank Wash. DC [online] Available from: https://gfdrr.org/sites/gfdrr.org/files/New%20Folder/Teisberg_EWS.pdf, 2009.

Thielen, J., Bartholmes, J., Ramos, M.-H. and Roo, A. de: The European flood alert system–Part 1: concept and development, Hydrol. Earth Syst. Sci., 13(2), 125–140, 2009.

Thiemann, M., Trosset, M., Gupta, H. and Sorooshian, S.: Bayesian recursive parameter estimation for hydrologic models Water Resources Research Volume 37, Issue 10, Water Resour. Res., 37(10), 2521–2535, doi:10.1029/2000WR900405, 2001.

Todini, E.: A mass conservative and water storage consistent variable parameter Muskingum-Cunge approach, Hydrol. Earth Syst. Sci., 11, 1645–1659, 2007.

Todini, E., Szollosi-Nagy, A. and Wood, E. F.: Adaptive state-parameter estimation algorithm for real time hydrologic forecasting; a case study, in In: IISA/WMO

Workshop on the Recent Developments in Real Time Forecasting /Control of Water Resources Systems, Laxemburg, Austria,., 1976.

Todini, E., Alberoni, P., Butts, M., Collier, C., Khatibi, R., Samuels, P. and Weerts, A.: ACTIF best practice paper–understanding and reducing uncertainty in flood forecasting, in P. Balabanis, D. Lumbroso, P. Samuels International conference on innovation, advances and implementation of flood forecasting technology, TromsØ, Norway., 2005.

Tokar, A. and Johnson, P.: Rainfall-Runoff Modeling Using Artificial Neural Networks, J. Hydrol. Eng., 4(3), 232–239, doi:10.1061/(ASCE)1084-0699(1999)4:3(232), 1999.

Tulloch, A. I. T. and Szabo, J. K.: A behavioural ecology approach to understand volunteer surveying for citizen science datasets, Emu, 112(4), 313, doi:10.1071/MU12009, 2012.

USGS: USGS discharge data for Riverside, Httpnwiswaterdatausgsgov Nwismonthly, 2016.

Vaché, K. B., McDonnell, J. J. and Bolte, J.: On the use of multiple criteria for a posteriori model rejection: Soft data to characterize model performance, Geophys. Res. Lett., 31(21), L21504, doi:10.1029/2004GL021577, 2004.

Valstar, J. R., McLaughlin, D. B., te Stroet, C. B. M. and van Geer, F. C.: A representer-based inverse method for groundwater flow and transport applications, Water Resour. Res., 40(5), W05116, doi:10.1029/2003WR002922, 2004.

Vandecasteele, A. and Devillers, R.: Improving volunteered geographic data quality using semantic similarity measurements, ISPRS-Int. Arch. Photogramm. Remote Sens. Spat. Inf. Sci., 1(1), 143–148, 2013.

Verlaan, M.: Efficient Kalman Filtering Algorithms for Hydrodynamic Models, PhD Thesis, Delft University of Technology, The Netherlands., 1998.

Verlaan, M. and Heemink, A. W.: Data assimilation schemes for non-linear shallow water flow models, Proc Second Int Symp Assim. Obs. Tokyo Jpn. WMO, 247–252, 1995.

Wagener, T. and Gupta, H. V.: Model identification for hydrological forecasting under uncertainty, Stoch. Environ. Res. Risk Assess., 19(6), 378–387, doi:10.1007/s00477-005-0006-5, 2005.

Walker, J. P. and Houser, P. R.: Hydrologic Data Assimilation, Adv. Water Sci. Methodol., 233–233, doi:10.5772/1112, 2005.

Walker, J. P., Willgoose, G. R. and Kalma, J. D.: One-dimensional soil moisture profile retrieval by assimilation of near-surface observations: a comparison of retrieval algorithms, Adv. Water Resour., 24(6), 631–650, doi:10.1016/S0309-1708(00)00043-9, 2001.

Weerts, A. H. and El Serafy, G. Y. H.: Particle filtering and ensemble Kalman filtering for state updating with hydrological conceptual rainfall-runoff models, Water Resour. Res., 42(9), 1–17, doi:10.1029/2005WR004093, 2006.

WeSenseIt: WeSenseIt: Citizen Water Observatories, [online] Available from: http://wesenseit.eu/ (Accessed 19 February 2016), 2016.

Westerberg, I. K., Guerrero, J. L., Younger, P. M., Beven, K. J., Seibert, J., Halldin, S., Freer, J. E. and Xu, C. Y.: Calibration of hydrological models using flow-duration curves, Hydrol. Earth Syst. Sci., 15(7), 2205–2227, doi:10.5194/hess-15-2205-2011, 2011.

Wheater, H. S., Jakeman, A. J. and Beven, K. J.: Progress and directions in rainfall-runoff modelling, Model. Change Environ. Syst. 1993 Pp 101-132, 101–132, 1993.

Whigham, P. A. and Crapper, P. F.: Modelling rainfall-runoff using genetic programming, Math. Comput. Model., 33(6–7), 707–721, doi:10.1016/S0895-7177(00)00274-0, 2001.

Wilby, R. L., Beven, K. J. and Reynard, N. S.: Climate change and fluvial flood risk in the UK: more of the same?, Hydrol. Process., 22(14), 2511–2523, doi:10.1002/hyp.6847, 2008.

WMO: Simulated real-time intercomparison of hydrological models, World Meteorological Organization., 1992.

Wood, S. J., Jones, D. A. and Moore, R. J.: Accuracy of rainfall measurement for scales of hydrological interest, Hydrol. Earth Syst. Sci. Discuss., 4(4), 531–543, 2000.

Xevi, E., Christiaens, K., Espino, A., Sewnandan, W., Mallants, D., Sørensen, H. and Feyen, J.: Calibration, Validation and Sensitivity Analysis of the MIKE-SHE Model Using the Neuenkirchen Catchment as Case Study, Water Resour. Manag., 11(3), 219–242, doi:10.1023/A:1007977521604, 1997.

Xie, X. and Zhang, D.: Data assimilation for distributed hydrological catchment modeling via ensemble Kalman filter, Adv. Water Resour., 33(6), 678–690, doi:10.1016/j.advwatres.2010.03.012, 2010.

Xuan, Y., Cluckie, I. D. and Wang, Y.: Uncertainty analysis of hydrological ensemble forecasts in a distributed model utilising short-range rainfall prediction, Hydrol. Earth Syst. Sci., 13(3), 293–303, 2009.

Yan, K., Baldassarre, G. D. and Solomatine, D. P.: Exploring the potential of SRTM topographic data for flood inundation modelling under uncertainty, J. Hydroinformatics, 15(3), 849–861, doi:10.2166/hydro.2013.137, 2013.

Yan, K., Di Baldassarre, G., Solomatine, D. P. and Schumann, G. J.-P.: A review of low-cost space-borne data for flood modelling: topography, flood extent and water level, Hydrol. Process., 29(15), 3368–3387, doi:10.1002/hyp.10449, 2015.

Yarvis, M., Kushalnagar, N., Singh, H., Rangarajan, A., Liu, Y. and Singh, S.: Exploiting heterogeneity in sensor networks, in Proceedings IEEE INFOCOM 2005. 24th Annual Joint Conference of the IEEE Computer and Communications Societies, vol. 2, pp. 878–890 vol. 2., 2005.

Young, P.: Recursive Estimation and Time-series Analysis: An Introduction, Springer-Verlag New York, Inc., New York, NY, USA., 1984.

Part of the cover page was designed by freepik.com

List of Acronyms

2D	Two Dimensional
3D	Three Dimensional
3D-Var	Three Dimensional variational assimilation
4DEnKF	Four Dimensional Ensemble Kalman Filter
4D-Var	Four Dimensional variational assimilation
AAWA	Alto Adriatico Water Authority
AEnKF	Asynchronous Ensemble Kalman filter
AEnKF	Asynchronous Ensemble Kalman Filter
AF	Analysis-forecast cycle
AMICO	Alto adriatico Modello Idrologico e idrauliCO
AMSR-E	Advanced Microwave Scanning Radiometer-EOS
ANN	Artificial neural networks
AP	Assimilation point
ARMA	Auto-regressive moving average
BBM	Black box model
CS	Crowdsourced
DA	Data Assimilation
DACO	DA approach for integrating Crowdsourced Observations
DDM	Data driven model
DEM	Digital Elevation Model
DFPMIN	Davidon-Fletcher-Powell minimization algorithm
DFW	Dallas–Fort Worth Metroplex area.
DHI	Danish hydraulic institute
DI	Direct insertion
DySc	Dynamic Social
ECWMF	European Centre for Medium-Range Weather Forecast

EFAS	European Flood Alert System
EFFS	European-scale flood forecasting system
EKF	Extended Kalman filter
EnKF	Ensemble Kalman filter
ErrRC	Uncertain in rating curve
ErrSD	Uncertain estimation of the synthetic discharge
ErrWL	Uncertain in WL estimation
EWS	Early Warning System
FGAT	First-Guess at the Appropriate Time
FI	Forecasted Input
FP	Framework Programme
GLUE	Generalized Likelihood Uncertainty Estimation
GR4J	Modèle du Génie Ruralà 4 paramètres Journalier
GRACE	Gravity Recovery and Climate Experiment
HBV	Hydrologiska Byras Vattenbalansavdelning
HEC-RAS	Hydrologic Engineering Center - River Analysis System
HYREX	Hydrological Radar Experiment Dataset
IMM	Interacting Multiple Model
IUH	Instantaneous Unit Hydrograph
KF	Kalman filter
KMN	Kalinin-Milyukov-Nash
MEL	Maximum citizen Engagement Level
MI	Measured input
MLEF	Maximum Likelihood Ensemble Filter
MMAE	Multiple model adaptive estimation
MODIS	Moderate Resolution Imaging Spectroradiometer
MS	Model structure
NNSE	Normalized Nash Sutcliffe index
NRR	Normalized RMSE Ratio
NS	Nudging scheme

NSE	Nash-Sutcliffe Efficiency
NWS	National Weather Service
ODA	Ocean data assimilation
PA	Ponte degli Angeli
PCR-GLOBWB	PCRaster GLOBal Water Balance model
PET	Potential evapotranspiration
PF	Particle filter
PM	Ponte Marchese
PT	Perturbation Time
QR	Quick Response
REnKF	Recursive ensemble Kalman filter
REW	Representative Elementary Watershed framework
RI	Repeated insertion
RMSD	Root Mean Square Difference
RMSE	Root Mean Square Error
SC	Spatial configuration
SCS-CN	Soil Conservation Service Curve Number
SME	Small Medium Enterprise
SRTM	Shuttle Radar Topography Mission
StPh	Static Physical
StSc	Static Social
TMI	TRMM Microwave Imager
TOF	Time Of Forecast
TRMM	Tropical Rainfall Measuring Mission
UKF	Unscented Kalman filter
WD	Water depth
WMO	World Meteorological Organization
WP	Work Package
WSI	WeSenseIt

LIST OF TABLE

List of Figures

ABOUT THE AUTHOR

Maurizio Mazzoleni was born in Brescia in November 1986. He graduated from University of Brescia, in Brescia, Italy, in May 2011. During his university studies he continued to pursue his interest in the flood protection by moving to UNESCO-IHE with the support of a scholarship awarded by University of Brescia to carry out his Master Thesis. Afterwards, he cooperated for 1 year within the KULTURisk Project as research fellow at the University of Brescia.

Currently, Mr. Mazzoleni is a PhD candidate at UNESCO-IHE Institute for Water Education under the Department of Integrated Water Systems and Governance, Delft, The Netherlands. His research interest include hydrologic and hydrodynamic modelling, in particular he dealt with issue related to flood forecasting, data assimilation, flood inundation mapping, flood risk and uncertainty analysis, flood defence systems design and reliability analysis, statistical hydrology.

Journals publications

- Mazzoleni M., Cortes Arevalo V.J., Wehn U., Alfonso L., and Solomatine D.P. (2016) Assimilation of crowdsourced observations into a cascade of hydrological and hydraulic models: The flood event of May 2013 in the Bacchiglione basin, Hydrology and Earth System Sciences, In preparation for Hydrology and Earth System Science
- Mazzoleni M., Chacon-Hurtado J., Noh S.J., Seo D.J., Alfonso L., and Solomatine D.P. (2016) Data assimilation in hydrologic routing: impact of sensor placement on flood prediction, Hydrological Processes, Hydrological Processes, under review
- Mazzoleni M., Noh S.J., Lee H., Liu Y., Seo D.J., and Solomatine D.P. (2016) Real-time assimilation of streamflow observations into a hydrologic routing model: Effects of different model structures and updating methods, Journal of Hydrology, under review

- Mazzoleni M., Veerlan M., Alfonso L., Monego M., Norbiato D., Ferri M., and Solomatine D.P. (2015) Can assimilation of crowdsourced streamflow observations in hydrological modelling improve flood prediction?, Hydrology and Earth System Sciences, under review
- Mazzoleni M., Alfonso L. and Solomatine D.P. (2015) Effect of spatial distribution and quality of sensors on the assimilation of distributed streamflow observations in hydrological modelling, Hydrological Sciences Journal, accepted
- Mazzoleni M., Dottori F., Brandimarte L., Tekle S. and Martina M. (2015) Effects of levee cover quality on flood mapping in case of levee breach due to overtopping, Hydrological Sciences Journal, accepted
- Mazzoleni M., Alfonso L., Chacon-Hurtado J.C. and Solomatine D.P. (2015) Assimilating uncertain, dynamic and intermittent streamflow observations in hydrological models", Advances in Water Resources, 83, 323-339,
- Mazzoleni, M., Barontini, S., Ranzi, R., and Brandimarte, L. (2014). Innovative Probabilistic Methodology for Evaluating the Reliability of Discrete Levee Reaches Owing to Piping, J. Hydrol. Eng., 10.1061/(ASCE)HE.1943-5584.0001055.
- Mazzoleni, M., Bacchi, B., Barontini, S., Di Baldassarre, G., Pilotti, M., and Ranzi, R. (2014). Flooding Hazard Mapping in Floodplain Areas Affected by Piping Breaches in the Po River, Italy, „1¤7„1¤7 J. Hydrol. Eng., 19(4), 717--731.

Conference proceedings

- Chacon-Hurtado J., Mazzoleni M., Alfonso L. and Solomatine D.P. (2016) Scheduling of dynamic hydrometric sensors for operational streamflow forecasting, Hydroinformatics Conference 2016, Incheon, Korea;
- Alfonso L., Chacon-Hurtado J., Mazzoleni M. and Solomatine D.P. (2016) Optimal Design of Hydrometric Monitoring Networks with Dynamic Components based on Information Theory, Hydroinformatics Conference 2016, Incheon, Korea;
- Chacon-Hurtado J., Mazzoleni M., Corzo G. and Solomatine D.P. (2016) On the use of surrogate inverse models for hydrological data assimilation, Hydroinformatics Conference 2016, Incheon, Korea;

- Mazzoleni M., Noh S.J., Lee H., Liu Y., Seo D.J. and Solomatine D.P. (2016) Assimilation of real-time streamflow observations into a hydrologic routing model: Effect of different model structures, Hydroinformatics Conference 2016, Incheon, Korea;
- Noh S.J., Mazzoleni M., Lee H., Liu Y., Seo D.J. and Solomatine D.P. (2016) Real-time assimilation of observations from heterogeneous sensors into hydrologic routing models, Hydroinformatics Conference 2016, Incheon, Korea;
- Viavattene C., McCarthy S., Ferri M., Monego M. and Mazzoleni M. (2016) Evaluation of emergency protocols using agent-based approach, 13th International Conference on Information Systems for Crisis Response and Management, Sao Paulo, Brazil
- Mazzoleni M., Chacon-Hurtado J., Alfonso Segura L. and Solomatine D.P. (2015) Towards the assimilation of anarchist flow observations in hydrological models, IAHR World Congress 2015, At Den Haag, The Netherlands
- Mazzoleni M., Barontini S., Ranzi R., and Brandimarte L. (2015) Effect of availability of levee data in the estimation of the probability of levee failure in case of piping, IAHR World Congress 2015, At Den Haag, The Netherlands
- Mazzoleni M., Alfonso Segura L. and Solomatine D. (2014) Effect of different hydrological model structures on the assimilation of distributed uncertain observations of discharge, Hydroinformatics Conference 2014, New York, USA;
- Mazzoleni M., Alfonso Segura L. and Solomatine D. (2014) Assimilation of heterogeneous uncertain data, having different observational errors, in hydrological models, Hydroinformatics Conference 2014, New York, USA;
- Mazzoleni M., Barontini S. and Ranzi R. (2012) Reliability levee model to support flooding hazard assessment, Proc. XXXIII Conference of Hydraulics and Hydraulic Engineering, Brescia (Italy), 10--14 September 2012, Bacchi B., Ranzi R. and Tomirotti M. (editors), ISBN: 978-88-97181-18-7 (on CD), Edibios, Cosenza (Italy), 10 pp;
- Ranzi R., Barontini S., Mazzoleni M., Ferri M. and Bacchi B. (2012) Levee breaches and "geotechnical uncertainty" in flood risk mapping, IAHR European Division Congress, Munich, 27--28 June 2012, Technische Universitat Munchen, 6 pp. (on USB Pen), 2012;

Netherlands Research School for the
Socio-Economic and Natural Sciences of the Environment

D I P L O M A

For specialised PhD training

The Netherlands Research School for the
Socio-Economic and Natural Sciences of the Environment
(SENSE) declares that

Maurizio Mazzoleni

born on 18 November 1986 in Brescia, Italy

has successfully fulfilled all requirements of the
Educational Programme of SENSE.

Delft, 28 November 2016

the Chairman of the SENSE board	the SENSE Director of Education
Prof. dr. Huub Rijnaarts	Dr. Ad van Dommelen

The SENSE Research School has been accredited by the Royal Netherlands Academy of Arts and Sciences (KNAW)

KONINKLIJKE NEDERLANDSE
AKADEMIE VAN WETENSCHAPPEN

The SENSE Research School declares that Mr Maurizio Mazzoleni has successfully fulfilled all requirements of the Educational PhD Programme of SENSE with a work load of 38.6 EC, including the following activities:

SENSE PhD Courses

- Environmental research in context (2014)
- Uncertainty propagation in spatial and environmental modelling, Wageningen University (2014)
- Research in context activity: 'Co-organising meeting of International Association for Hydro-Environment Engineering and Research - Young Professional Network Delft (IAHR -YPN)' (2015)

Other PhD and Advanced MSc Courses

- MSc course Data-driven modelling and real-time control of water systems, UNESCO-IHE Delft (2012)

External training at a foreign research institute

- Research visit on 'The effect of various model updating techniques in the assimilation of streamflow observations in hydrologic routing modelling, University of Texas, United States (2015)

Management and Didactic Skills Training

- Teacher assistant in the BSc courses 'River structures', 'Introduction to data assimilation', 'Numerical methods', 'Mathematical Formulation of fluid flow equations', 'Hydroinformatics modelling in MATLAB' (2013-2016)
- Teaching MSc seminar 'Probabilistic design of levee system', Asociación Peruana de Ingeniería Hidráulica y Ambiental (APIHA), Peru (2014)
- Supervising four MSc students at UNESCO-IHE Delft (2014-2015)
- Co-organising LaTeX - introductory course, UNESCO-IHE Delft and International Association for Hydro-Environment Engineering and Research (IAHR-YPN) (2015)
- Co-organising and convening the workshop 'Hydroinformatics for hydrology: introduction to data science including data assimilation', European Geosciences Union General Assembly, Vienna, Austria (2016)

Selection of Oral Presentations

- *Effect of different hydrological model structures on the assimilation of distributed uncertain observations of discharge.* 11th International Conference on Hydroinformatics (HIC2014), 17-21 August 2014, New York, United States
- *Towards the assimilation of anarchist flow observations in hydrological models.* 36th IAHR World Congress, 28 June-3 July 2015, Delft-The Hague, The Netherlands
- *Comparative analysis of various real-time data assimilation approaches for assimilating streamflow into a hydrologic routing model.* European Geosciences Union General Assembly (EGU2016), 17-22 April 2016, Vienna, Austria
- *Towards real-time assimilation of crowdsourced observations in hydrological modelling.* European Geosciences Union General Assembly (EGU2016), 17-22 April 2016, Vienna, Austria

SENSE Coordinator PhD Education

Dr. ing. Monique Gulickx

For Product Safety Concerns and Information please contact our EU representative GPSR@taylorandfrancis.com Taylor & Francis Verlag GmbH, Kaufingerstraße 24, 80331 München, Germany

Printed and bound by CPI Group (UK) Ltd, Croydon, CR0 4YY
08/05/2025
01864478-0001